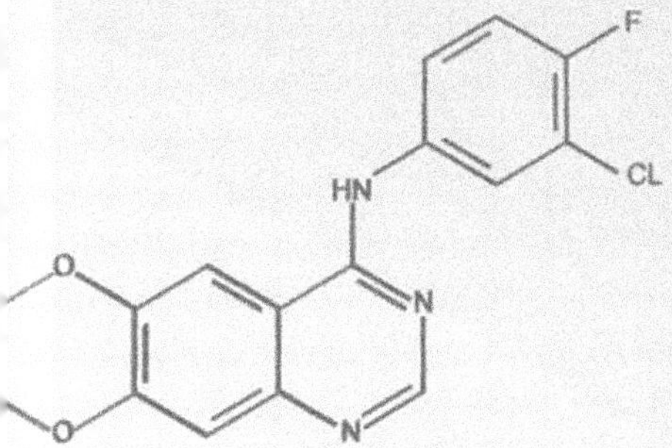

Cancer Genetics and Genomics for Personalized Medicine

Cancer Genetics and Genomics for Personalized Medicine

edited by

Il-Jin Kim

PAN STANFORD PUBLISHING

Published by

Pan Stanford Publishing Pte. Ltd.
Penthouse Level, Suntec Tower 3
8 Temasek Boulevard
Singapore 038988

Email: editorial@panstanford.com
Web: www.panstanford.com

British Library Cataloguing-in-Publication Data
A catalogue record for this book is available from the British Library.

Cancer Genetics and Genomics for Personalized Medicine

ISBN 978-981-4669-87-0 (Hardcover)
ISBN 978-1-315-36476-6 (eBook)

Printed in Great Britain by Ashford Colour Press Ltd

FSC www.fsc.org MIX Paper from responsible sources FSC® C011748

Contents

Preface

On a busy morning one day, I got an email from a publisher asking for a new publication on precision and personalized medicine for cancer treatment and research. By that time, I had led several relevant projects, including the development of new next-generation sequencing (NGS) technologies and corresponding bioinformatics programs. There were surely successful and commercial products developed from these projects (which are now available in the genetic analysis market in the world), yet I was not sure if I would be able to write or edit a textbook on a big topic like personalized medicine. Previously, I only had had a chance to join two scientific books as a single-chapter contributor, but had never written or edited a whole book by myself. It seemed quite overwhelming to me at the time. However, as several months passed since I received the first email about this opportunity, I began to open my mind and be more positive towards the idea. In fact, I had been working on personalized medicine, cancer genetics, companion diagnostics, and cancer biomarker discovery for more than 15 years by then. Thus, I thought it would be good to organize and integrate all the knowledge that I have along with that which top-level scientists, researchers, and medical doctors have been gathering regarding personalized medicine in cancer. Consequently, I gladly decided to embark on this project even though I was very busy developing multiple new sequencing and genetic technologies. First, I tried to select the topics that could be the most beneficial for the people engaged in training such as graduate students, medical residents, and other high-level professionals who are relatively unfamiliar with personalized medicine. The next thing I did was the most time-consuming and challenging work—finding the most suitable people who can write on the topic with up-to-date knowledge and information in a plain language. As a matter of fact, it took me almost an entire year to find all the world-class authors for each chapter. I really appreciate all the contributors of each chapter in this book. Without their contribution and efforts, the book could not have been published. I thank them for their patience and support because it took much longer than expected to finally get this book published. I also appreciate my good friend and colleague Pedro Mendez, who designed the cover for this book. I must say that he is the most artistic person I have ever met in my group. I also thank James Kim for his editorial assistance and sincere friendship. I thank the whole team at the UCSF Thoracic Oncology Program and CureSeq for their support and inspiration on precision medicine and companion diagnostics in developing new weapons (the world's fastest mutation screening assay) fighting cancer. Finally, I thank Stanford Chong greatly for suggesting and giving me a chance to publish this book, and Sarabjeet Garcha for all the help and communication.

I would like to offer my wholehearted gratitude to my family who supported me not only in the writing of this book but also in all my work in general. My parents, Ho-Young Kim and Yong-Soon Bang, in Korea taught me well to maintain a strong passion for my life's goals, and also to give endless love to my family. My wife, Hio Chung Kang, and my two sons, Thomas Kunhee and Benjamin Kunjune Kim, are truly the very source of all my sweat, effort, motivation, and energy behind the achievements.

We are already in the era of personalized and precision medicine for curing cancer. Technologies and methods will develop continuously and evolve rapidly. However, grasping the core concept and principles will constantly remain crucial until we finally cure cancer. I really hope this book will be helpful and informative to current and future heroes and heroines in their fight against, and the eventual conquest over, cancer.

Il-Jin Kim
February 2017
San Francisco

Chapter 1

Personalized Medicine for Cancer: Introduction and Overview of the Book

Il-Jin Kim and David Jablons
Thoracic Oncology Laboratory, Department of Surgery, Comprehensive Cancer Center, University of California San Francisco, San Francisco, CA, USA
Il-Jin.Kim@ucsf.edu, ijkimsf@gmail.com

1.1 Changing the Treatment Paradigm for Cancer

We live in an age in which tremendous technological breakthroughs in many fields—the Internet, mobile computers and smart phones, new energy technologies, to name just a few—have dramatically changed the global economy and human life. However, treatments for human disease have not kept pace with advances in other fields. One of the very few examples is antiretroviral therapy for HIV, whereby multiple-drug treatment can dramatically delay disease progression and prolong survival so that AIDS becomes a chronic disease [1].

For cancer, the second-most lethal human disease in the United States [2], there has been no significant improvement for decades. The reasons include continued high smoking rates, air pollution,

Cancer Genetics and Genomics for Personalized Medicine
Edited by Il-Jin Kim

ISBN 978-981-4669-87-0 (Hardcover), 978-1-315-36476-6 (eBook)
www.panstanford.com

exposure to environmental toxic material (e.g., asbestos), and many unknown epidemiological and genetic causes. Cancer is known to be a genetic disease mainly caused by genetic alterations such as somatic mutations of tumor suppressor genes (e.g., TP53) and oncogenes (e.g., K-ras and EGFR) [3, 4]. However, cancer is notoriously complex and heterogeneous. In addition to genetic alterations, many other mechanisms, such as epigenetic, immune related, and environmental factors, contribute to cancer development, evolution, metastasis, and acquisition of drug resistance [3, 4]. Moreover, cancer is not static—its characteristics change to survive in different situations and environments in the human body. Thus, it is not easy to identify a clear target to attack and cure cancer.

Among all potential factors that cause a high incidence of cancer and inhibit effective treatment, the biggest culprit may be the "one-fits-all" approach. This classical approach has been used since the first chemotherapy treatment was given in the 1940s [5], but oncologists and cancer researchers have known for years that it is rarely successful. A new paradigm of cancer treatment has been needed to move into the modern era of "personalized" or "precision" medicine (PM), which can be defined as a medical treatment decision and action based on individual patient's genetic, epigenetic, histopathological, or other health information.

The first interruption of the one-fits-all treatment approach was the identification of BCR-ABL fusion protein in patients with chronic myelogenous leukemia (CML) [6, 7]. Soon after this discovery, imatinib (Gleevec) was developed to cure CML with BCR-ABL fusion. This was to become the first successful PM attempt. After imatinib treatment for BCR-ABL positive CML patients, the 5-year survival rate has doubled from 31% in the early 1990s to 60% in 2010 [8]. The FDA later approved an imatinib treatment for KIT-positive gastrointestinal stromal tumors (GIST) patients [9, 10] and Philadelphia chromosome-positive acute lymphoblastic leukemia (Ph+ ALL) patients [10]. The imatinib breakthrough has also shifted the research focus for cancer treatment in many academic institutes and pharmaceutical companies that now seek to identify molecular markers and drug targets. For example, hormone epidermal growth factor receptor 2 (HER-2), estrogen receptors, and progesterone receptors were identified in breast cancer patients

[11], and trastuzumab (Herceptin) has proved effective for those with HER-2 positive tumors [11].

For lung cancer, EGFR mutations and EML4-ALK fusion proteins have been identified and patients with EGFR oncogenic mutations were found to be highly responsive to tyrosine kinase inhibitor (TKI), gefitinib, or erlotinib. This entailed the new concept of routine screening of EGFR mutations in lung cancer patients to find the subpopulation suitable for TKI treatment. However, many of these patients eventually experience tumor recurrence because of drug resistance caused by EGFR T790M mutation, MET amplification, noncoding RNA-related mechanisms, or transformation of the tumor's identity [12]. A recently identified fusion protein, EML4-ALK, in a lung cancer patient harboring no EGFR and K-ras mutation, may be a new target [13]. Interestingly, the ALK inhibitor crizotinib was not a successful cancer treatment until this new oncogenic fusion was identified. It took less than 5 years from identifying the new marker, EML4-ALK, to FDA approval for crizotinib for lung cancer patients. This may be the fastest approval on record for a cancer drug and illustrates a new paradigm of "drug repurposing or repositioning" after new molecular markers are identified.

The focus of this book is on the genetic and genomic aspects of PM for cancer. Dr. Daniela Morales-Espinosa et al. detail a PM approach for treating lung cancer patients in Chapter 2, and Drs. Jae-Jun Shim and Ju-Seog Lee describe PM for liver cancer in Chapter 3.

1.2 Companion Diagnostics and New Sequencing Technologies

As molecular markers like EGFR mutations and targeting drugs like gefitinib initiated a new era of PM, robust and reliable methods are needed to detect molecular alterations that identify patient subpopulations suitable for a new treatment. Companion diagnostics can be defined as "an assay or diagnostic system for selecting and identifying the best treatment decision for patients" [14]. One example is the fluorescence in situ hybridization (FISH) assay developed by Abbott Diagnostics for selecting lung cancer patients with EML4-ALK. Developing a companion diagnostic assay is tremendously expensive, but an accurate and cost-effective assay is key to the PM approach.

Next-generation sequencing (NGS) has become popular for genetic analysis of mutation (exome or whole genome sequencing [WGS]) and RNA expression (transcriptome or RNA-seq) analysis. NGS uses either a hybridized capturing technology (Illumina HiSeq and MiSeq) or a semiconductor sequencing technology that detects hydrogen ions released during DNA polymerization (e.g., ion torrent PGM and proton). These methods can scan a large region (e.g., exome or WGS) or screen a specific target region (e.g., targeted panel for companion diagnostics). The NGS method that we developed, called NextDay Seq (NDS), enables fast, targeted sequencing of clinically actionable genes such as EGFR and K-ras in lung, colorectal, and other cancers (these technologies are licensed from UCSF to CureSeq Inc. www.cureseq.com). Sequencing can be completed within 48 hours from a DNA extraction of formalin-fixed paraffin-embedded (FFPE) samples through NDS to final data analysis. A targeted NGS panel can be used to discover a companion diagnostic tool for efficiently selecting specific types of cancer for targeted therapy.

NGS still needs to be improved. Although exome or whole genome screening is powerful, it is not trivial to filter out many false positives and accurately select true mutations. A homopolymer or a repeat of the same sequence is one major problem for NGS. As many microsatellite and loss of heterozygosity (LOH) markers are just repeats of the same sequences, it can be an issue to use NGS for studying those markers. For example, around 10% of colorectal cancer samples have a microsatellite instability caused by a deletion or insertion of microsatellite sequences [16]. Thus, it is important to select the best NGS method for a particular type of genetic screening. This is also illustrated by the following example of EGFR. Currently, EGFR may be the most important gene for PM in lung cancer. Using different NGS technologies to screen more than 1000 lung cancer tissues, we found that some technologies have a problem detecting EGFR exon 19 deletion mutations. Unlike other common oncogenic mutations (i.e., EGFR L858R, K-ras codon 12 and 13 mutations, and BRAF V600E), EGFR exon 19 deletions are multiple bp deletions as long as 15–25 bp. This kind of long deletion can cause a wrong sequence alignment or be regarded as sequence noise by bioinformatics programs. Next on the horizon is third-generation sequencing or nanopore sequencing, which can sequence up to 50 kb [17] and might overcome several issues found in NGS.

1.3 Early Detection of Cancer and Tumor Recurrence Monitoring: Circulating Tumor Cell (CTC) and Circulating Tumor DNA (ctDNA)

Although new drug targets and molecular markers for novel, efficient therapy are needed to ultimately reduce the high mortality rates for cancer, a more practical approach will be to detect cancer cells at an early stage and provide the best treatment option, such as a surgical resection or targeted therapy if available.

Cancer can be detected early by imaging methods, such as X-ray, computerized tomography, or magnetic resonance imaging, and by less invasive or non-invasive screening of blood, sputum, urine, stool, and exhaled breath condensate [19, 20]. Currently, imaging methods may be a golden standard, but they may entail low sensitivity and specificity, high cost, and inconvenient access to an up-to-date facility.

Molecular screening could be a promising approach for detecting cancers early, but is in the preliminary stage and needs considerable clinical validation. To use human body fluids or samples for early cancer diagnosis, two issues should be solved. The first is to identify reliable molecular markers. Despite many studies that tested early detection of cancer by mutation, methylation, and noncoding RNA using blood, sputum, and other biological samples [19, 20], only a few methods were approved for early cancer screening. One example is prostate cancer antigen 3 (PCA3), which is a long noncoding RNA expressed mainly in prostate cancer [20, 21]. PCA3 is FDA-approved for a urine test for prostate cancer screening using real-time quantitative PCR [20, 21]. The second issue is to use a reliable and highly sensitive technology to detect cancer early. The ability to detect cancer-specific mutations such as TP53, K-ras, and EGFR in plasma or sera came about when very sensitive technologies such as NGS became available [19–22]. Circulating tumor DNA (ctDNA) and circulating tumor cells (CTC) are promising molecular markers for the early detection of cancer [22, 23]. Cell-free fetal DNA is clinically used for non-invasive prenatal testing (NIPT) by a mass spectrometry-based method, digital PCR, or NGS [24]. While it is relatively common to use cfDNA for NIPT, using ctDNA analysis to detect cancer is in the very early stage, though results are encouraging [22, 23]. Unlike ctDNA, CTC was FDA-approved for metastatic breast, colorectal, or prostate cancer screening with CellSearch, an epithelial cell adhesion model (EpCAM)-based method [25, 26]. Although several issues have been raised for CTC screening, such as a lack of metastasis detection and non-epithelial cell originated cancer detection [25, 26], it is encouraging that a less invasive blood-based screening tool is currently available for cancer screening in the clinic.

In Chapter 4, Drs. Dana Tsui and Muhammed Murtaza relate how ctDNA has been used for PM in cancer research and treatment. In a Chapter 5, Dr. Jin Sun Lee et al. describe the use of CTC in breast cancer research and treatment.

1.4 Cancer Animal (Mouse) Models and Microenvironment for Personalized Medicine

Many promising drug candidates showing striking tumor-inhibiting effects in in vitro ultimately fail in clinical trials. Although cancer cell lines with mutations (i.e., K-ras, EGFR, or BRAF) and human cancer tissues are wonderful resources for research, they are clearly limited in their ability to reflect real human immune, angiogenesis, microenvironment, and other physiological conditions. Thus, any new drug target or biological marker requires testing in an in vivo system. In Chapter 6, Dr. Laura Soucek and her colleagues comprehensively review an "avatar" mouse model of human cancer. A fast and reliable human cancer avatar model would be very helpful for selecting the best treatment regimen for cancer patients.

Drug resistance is a major obstacle for current and future PM therapy. The mechanisms of drug resistance can be genetic alterations such as EGFR T790M mutation and MET amplification, in the case of TKI treatment, and acquired ALK point mutations in the case of crizotinib treatment. In addition to the genetic alterations for drug resistance, Dr. Yu Sun explains in Chapter 7 that the major reason for drug resistance is an interaction between cancer and microenvironment. Because cancer is evolving and escaping from an attack by the human immune system, the cancer microenvironment must be understood in order for drug resistance to be overcome.

1.5 Personalized Immunotherapy

Although this book is mainly focused on the genetic and genomic aspects of PM for cancer, we also consider personalized immunotherapy as a future treatment. As cancer cells grow and treatment starts, cancer cells are recognized by the therapeutic agent (either chemotherapy or targeted therapy) or our immune system. But cancer cells can change their identity and escape from attack by a drug or the immune system. If the immune system can be boosted to efficiently kill only tumor cells, it can be very specific and effective. Dr. Joost Hegmans et al. describe a cellular personalized immunotherapy in Chapter 8, and Dr. G.E. Bates et al. describe another PM treatment, hyperthermic intraperitoneal chemotherapy (HIPEC), in Chapter 9.

1.6 Hereditary Cancer Syndromes and Potential Treatment

Many human organs have both sporadic and hereditary cancer syndromes. A challenging application of PM approaches would be for hereditary cancers. Examples include hereditary breast and ovarian cancer, and many hereditary syndromes of gastrointestinal organs (hereditary nonpolyposis colorectal cancer, familial adenomatous polyposis, juvenile polyposis, Peutz–Jeghers syndrome, and more). Most of these syndromes are caused by genetic alterations of genes, such as BRCA1/2 for hereditary breast and ovarian cancer, and mismatch repair genes for hereditary nonpolyposis colorectal cancer. While many hereditary cancer syndromes are caused by autosomal dominant inheritance, some, such as MYH-associated polyposis, are caused by an autosomal recessive inheritance [27].

Hereditary cancers account for 1–5% of all human cancers and each syndrome has its own selection criteria and clinical surveillance. For example, surgical resection is recommended for an APC germline mutation carrier in families with familial adenomatous polyposis, and for a BRCA1 or BRCA2 germline mutation carrier. However, surgery is not always feasible after these genes are identified. Moreover, the causing genes are not known for every hereditary cancer syndrome, making screening family members for early mutation impractical. Thus, it is still very challenging to provide optimal therapy to hereditary cancer patients. In Chapter 10, Dr. Vivek Subbiah provides an overview and perspective for the PM treatment of hereditary cancer patients. If genetic screening tools like NGS can be used to develop suitable cures, it would dramatically change the surveillance of hereditary cancer patients.

1.7 Future Directions

This book has recounted how recent progress in the field of PM has yielded. To develop comprehensive treatments and cures for patients with cancer, we believe the field will need to make progress in the following areas.

(1) Identification of clinically applicable therapeutic targets and biomarkers The concept of PM seems to be widely accepted in research and clinical domains. There is no question that the "one-fits-all" treatment strategy does not work for curing cancer or other diseases. Although tremendous efforts such as The Cancer Genome Atlas and International Cancer Genome Consortium have sought to identify molecular and genetic alterations, the number of clinically actionable drug targets and biomarkers remains limited. K-ras and TP53 are among the most frequently mutated genes in human cancers, but there is no effective therapy that targets them. For lung cancer specifically, a PM approach has thus far been successful only for adenocarcinoma, and fewer than half of patients with lung adenocarcinoma can benefit from targeted therapy with TKI (patients with EGFR activating mutation), BRAF inhibitor (patients with BRAF V600E mutation), and crizotinib (patients with EML4-ALK fusion). Although these are all truly impressive targeted treatments, for the many patients with lung adenocarcinoma or other types of lung cancer, other actionable alterations and targeting drugs are needed. Developing more clinically meaningful biomarkers and therapeutic targets for PM will likely require a comprehensive and integrative systems biology approach that combines genetics, epigenetics, metabolomics, and immunity, among other disciplines. A comprehensive review of pathology in PM era is described in Chapter 11. Noncoding RNA is also described as a promising PM biomarker and therapeutic target in Chapter 12. An application of pharmacokinetics for PM is described in Chapter 13.

(2) A patient monitoring system to overcome drug resistance
Almost every currently available anticancer drug quickly develops resistance by various mechanisms. Thus, although targeted therapy may benefit a specific cancer subgroup, it may not prolong survival significantly. A sequencing approach for ctDNA in plasma samples of cancer patients can be a good candidate for monitoring disease progression [22, 28, 29]. Exome sequencing, whole genome sequencing, and other approaches can be used to identify the recurrence mechanism. A system for monitoring patients after drug treatment will be needed to track tumor status and drug resistance mechanisms in order to select the next therapy.

(3) Standardization of biological sample collection and processing
Although there is some consensus about selecting biomarkers (e.g., EGFR, BRAF, and EML4-ALK), methods (i.e., NGS, real-time qPCR, and FISH), and treatment option, there is little, if any, consensus about collecting and handling patient samples. Blood samples would be one of the best less invasive biological specimens for many PM approaches. Circulating tumor cells and ctDNA hold great promise for PM, as described in Chapters 4 and 5, but it is not clear which samples are better (plasma vs. serum), which collecting tubes are

better (EDTA or heparinized or special, chemical treated), or what kind of sample preparation methods are preferable. A detailed discussion and consensus are needed internationally on the part of academic and industrial parties in order to guarantee highly reproducible results that would benefit cancer patients.

(4) Less expensive and complicated technologies Single-cell analysis is now available [30] and third-generation nanopore sequencing [17] may become a standard in PM. However, such technological advances tend to be too complicated and expensive to handle in a regular clinic. Currently, NGS is a useful and powerful technology for identifying (biomarker discovery and mechanism identification) and screening (finding known key mutations) genetic alterations, but most small hospitals and clinics are not equipped to use NGS. It requires an expensive machine and bioinformatics analysis skills. Although a Clinical Laboratory Improvement Amendments (CLIA)-certified lab can be used, cost and turnaround time are still big issues, as are accuracy and interpretation of the data. Therefore, a simple, fast, and cost-effective genetic analysis system for PM will need to be developed so that PM would be available not only in resource-rich countries.

(5) Data sharing and standardizing how patient information is stored The "N of 1" approach is to conduct a clinical trial in one individual instead of a large group of patients. It is ideal to test drug efficacy individually and select the best treatment regimen for each person. Although the N of 1 approach can be an ultimate goal for PM, it currently takes too much time and effort. Therefore, an intermediate approach may be appropriate in which a subpopulation and a specific group of diseases sharing the same or similar genetic genotype and biological phenotype are identified. The US National Institutes of Health Genomic Data Sharing policy [31], which became effective in January 2015, will encourage meaningful clinical and genetic data sharing. However, usage of different data and information-processing systems will inhibit a comprehensive and integrative analysis of large human genome data. Therefore, we should also standardize the process of storing patient information and sharing it among multiple organizations and nations (ideally regardless of the funding source). Finally, since many genetic variations show ethnic differences (i.e., high EGFR mutation frequency in young Asian female nonsmokers), specific consortiums may be useful among organizations and nations where similar genetic patterns are prominent.

(6) Co-developing and sharing companion diagnostic tools Although it will likely be considered controversial by industry partners in the field of PM, co-development is required to bring a robust companion diagnostic tool, in addition to the tested drug, to the market, as is required for FDA approval. It can easily cost millions of dollars to develop a new companion diagnostic tool such as FISH or NGS for PM. Interestingly, many major pharmaceutical companies with new potentially good drugs withhold an application for FDA approval due to the tremendous cost of developing a companion diagnostic tool. Reaching a consensus and building a consortium among different parties in order to developing companion diagnostic tools will not be easy, but would maximize resources, save time and costs, and ultimately lead to an effective PM approach for many diseases.

References

1. https://aidsinfo.nih.gov/guidelines/html/1/adult-and-adolescent-arv-guidelines/0
2. http://www.cdc.gov/nchs/fastats/leading-causes-of-death.htm
3. Hanahan D, Weinberg RA. The hallmarks of cancer. *Cell.* 2000 Jan 7;100(1):57–70.
4. Hanahan D, Weinberg RA. Hallmarks of cancer: the next generation. *Cell.* 2011 Mar 4;144(5):646–74.
5. DeVita VT Jr, Chu E. A history of cancer chemotherapy. *Cancer Res.* 2008 Nov 1;68(21):8643–53.
6. Shtivelman E, Lifshitz B, Gale RP, Roe BA, Canaani E. Alternative splicing of RNAs transcribed from the human abl gene and from the bcr-abl fused gene. *Cell.* 1986 Oct 24;47(2):277–84.
7. Ben-Neriah Y, Daley GQ, Mes-Masson AM, Witte ON, Baltimore D. The chronic myelogenous leukemia-specific P210 protein is the product of the bcr/abl hybrid gene. *Science.* 1986 Jul 11;233(4760):212–4.

8. http://www.cancer.net/cancer-types/leukemia-chronic-myeloid-cml/statistics
9. Heinrich MC, Blanke CD, Druker BJ, Corless CL. Inhibition of KIT tyrosine kinase activity: a novel molecular approach to the treatment of KIT-positive malignancies. *J Clin Oncol.* 2002 Mar 15;20(6):1692–703.
10. http://www.cancer.gov/about-cancer/treatment/drugs/imatinibmesylate
11. Foulkes WD, Smith IE, Reis-Filho JS. Triple-negative breast cancer. *N Engl J Med.* 2010 Nov 11;363(20):1938–8.
12. Oser MG, Niederst MJ, Sequist LV, Engelman JA. Transformation from non-small-cell lung cancer to small-cell lung cancer: molecular drivers and cells of origin. *Lancet Oncol.* 2015 Apr;16(4):e165–72. Review.
13. Soda M, Choi YL, Enomoto M, Takada S, Yamashita Y, Ishikawa S, Fujiwara S, Watanabe H, Kurashina K, Hatanaka H, Bando M, Ohno S, Ishikawa Y, Aburatani H, Niki T, Sohara Y, Sugiyama Y, Mano H. Identification of the transforming EML4-ALK fusion gene in non-small-cell lung cancer. *Nature.* 2007 Aug 2;448(7153):561–6.
14. Papadopoulos N, Kinzler KW, Vogelstein B. The role of companion diagnostics in the development and use of mutation-targeted cancer therapies. *Nat Biotechnol.* 2006 Aug;24(8):985–95.
15. Sanger F, Coulson AR. A rapid method for determining sequences in DNA by primed synthesis with DNA polymerase. *J Mol Biol.* **94** (3): 441–8. 1975.
16. Kim IJ, Shin Y, Kang HC, Park JH, Ku JL, Park HW, Park HR, Lim SB, Jeong SY, Kim WH, Park JG. Robust microsatellite instability (MSI) analysis by denaturing high-performance liquid chromatography (DHPLC). *J Hum Genet.* 2003;48(10):525–30.
17. Wang Y, Yang Q, Wang Z. The evolution of nanopore sequencing. *Front Genet.* 2015 Jan 7;5:449.
18. Gessner C, Kuhn H, Toepfer K, Hammerschmidt S, Schauer J, Wirtz H. Detection of p53 gene mutations in exhaled breath condensate of non-small cell lung cancer patients. *Lung Cancer.* 2004 Feb;43(2):215–22.
19. Hubers AJ, Prinsen CF, Sozzi G, Witte BI, Thunnissen E. Molecular sputum analysis for the diagnosis of lung cancer. *Br J Cancer.* 2013 Aug 6;109(3):530–7.
20. de Kok JB, Verhaegh GW, Roelofs RW, Hessels D, Kiemeney LA, Aalders TW, Swinkels DW, Schalken JA. DD3(PCA3), a very sensitive and specific marker to detect prostate tumors. *Cancer Res.* 2002 May 1;62(9):2695–8.

21. Prensner JR, Chinnaiyan AM. The emergence of lncRNAs in cancer biology. *Cancer Discov.* 2011 Oct;1(5):391–407.
22. Murtaza M, Dawson SJ, Tsui DW, Gale D, Forshew T, Piskorz AM, Parkinson C, Chin SF, Kingsbury Z, Wong AS, Marass F, Humphray S, Hadfield J, Bentley D, Chin TM, Brenton JD, Caldas C, Rosenfeld N. Non-invasive analysis of acquired resistance to cancer therapy by sequencing of plasma DNA. *Nature.* 2013 May 2;497(7447):108–12.
23. Heitzer E, Ulz P, Geigl JB. Circulating tumor DNA as a liquid biopsy for cancer. *Clin Chem.* 2015 Jan;61(1):112–23.
24. Bianchi DW, Parker RL, Wentworth J, Madankumar R, Saffer C, Das AF, Craig JA, Chudova DI, Devers PL, Jones KW, Oliver K, Rava RP, Sehnert AJ; CARE Study Group. DNA sequencing versus standard prenatal aneuploidy screening. *N Engl J Med.* 2014 Feb 27;370(9):799–808.
25. Ignatiadis M, Riethdorf S, Bidard FC, Vaucher I, Khazour M, Rothé F, Metallo J, Rouas G, Payne RE, Coombes R, Teufel I, Andergassen U, Apostolaki S, Politaki E, Mavroudis D, Bessi S, Pestrin M, Di Leo A, Campion M, Reinholz M, Perez E, Piccart M, Borgen E, Naume B, Jimenez J, Aura C, Zorzino L, Cassatella M, Sandri M, Mostert B, Sleijfer S, Kraan J, Janni W, Fehm T, Rack B, Terstappen L, Repollet M, Pierga JY, Miller C, Sotiriou C, Michiels S, Pantel K. International study on inter-reader variability for circulating tumor cells in breast cancer. *Breast Cancer Res.* 2014 Apr 23;16(2):R43.
26. Gorges TM, Pantel K. Circulating tumor cells as therapy-related biomarkers in cancer patients. *Cancer Immunol Immunother.* 2013 May;62(5):931–9.
27. Kim DW, Kim IJ, Kang HC, Jang SG, Kim K, Yoon HJ, Ahn SA, Han SY, Hong SH, Hwang JA, Sohn DK, Jeong SY, Choi HS, Hong CW, Lim SB, Park JG. Germline mutations of the MYH gene in Korean patients with multiple colorectal adenomas. *Int J Colorectal Dis.* 2007 Oct;22(10):1173–8.
28. Olsson E, Winter C, George A, Chen Y, Howlin J, Tang MH, Dahlgren M, Schulz R, Grabau D, van Westen D, Fernö M, Ingvar C, Rose C, Bendahl PO, Rydén L, Borg Å, Gruvberger-Saal SK, Jernström H, Saal LH. Serial monitoring of circulating tumor DNA in patients with primary breast cancer for detection of occult metastatic disease. *EMBO Mol Med.* 2015 May 18;7(8):1034–47.
29. Forshew T, Murtaza M, Parkinson C, Gale D, Tsui DW, Kaper F, Dawson SJ, Piskorz AM, Jimenez-Linan M, Bentley D, Hadfield J, May AP, Caldas C, Brenton JD, Rosenfeld N. Noninvasive identification and monitoring of cancer mutations by targeted deep sequencing of plasma DNA. *Sci Transl Med.* 2012 May 30;4(136):136ra68.

30. Navin N, Kendall J, Troge J, Andrews P, Rodgers L, McIndoo J, Cook K, Stepansky A, Levy D, Esposito D, Muthuswamy L, Krasnitz A, McCombie WR, Hicks J, Wigler M. Tumour evolution inferred by single-cell sequencing. *Nature.* 2011 Apr 7;472(7341):90–4.
31. http://gds.nih.gov/03policy2.html

Chapter 2

Personalized Medicine in Lung Cancer

Daniela Morales-Espinosa,[a] Silvia García-Román,[b] and Rafael Rosell[a,c]

[a] *Instituto Oncológico, Quirón-Dexeus University Hospital, C. Sabino Arana 5-19, Consulta-1.1, 08028 Barcelona, Spain*
[b] *Institut Químic Sarriá, Via Augusta, 390, 08017 Barcelona, Spain*
[c] *Catalan Institute of Oncology, Germans Trias i Pujol Health Sciences Institute and Hospital, Autonomous University of Barcelona (UAB), Campus Can Ruti, 08916 Badalona, Barcelona, Spain*
rrosell@iconcologia.net

2.1 Introduction

Lung cancer remains the leading cause of cancer-related death worldwide.[1] Most non-small-cell lung cancer (NSCLC) patients are diagnosed at advanced stages, when chemotherapy is generally considered palliative treatment. In recent years, such treatment has proved not only to prolong survival but also to improve quality of life (including reduction of disease-related symptoms) for these patients.[2] Even today, after all the advances in therapeutic oncology, failure to therapy and sometimes extensive side effects remain a challenge. Due to growing knowledge, the generalized, non-individualized, mechanistic use of very expensive treatments is

Cancer Genetics and Genomics for Personalized Medicine
Edited by Il-Jin Kim

ISBN 978-981-4669-87-0 (Hardcover), 978-1-315-36476-6 (eBook)
www.panstanford.com

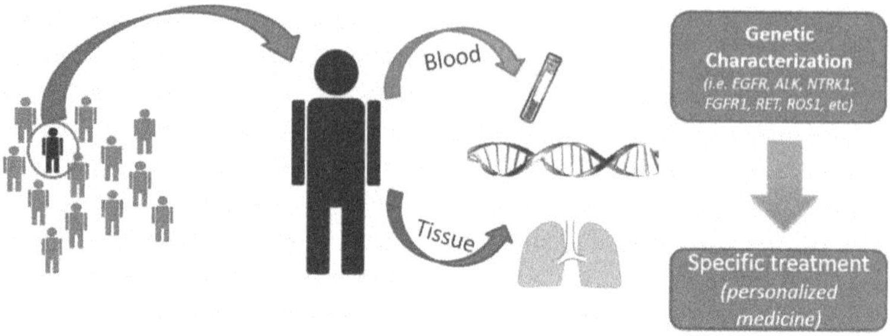

Figure 2.1 Personalized treatment approach.

becoming less acceptable over time. The pursuit of specific targets as an effort to personalize treatment is the preferred therapeutic approach (Fig. 2.1).

2.1.1 Predictive Models

There are huge differences in recurrence and survival rates among patients in the same clinicopathologic group, indicating that current treatment approaches result in overtreatment of some patients and undertreatment of others.[3,4]

Recently, molecular expression profiling of the tumor has emerged as a powerful prognostic tool to predict long-term outcomes in many cancers, including breast,[5] colon,[6] and liver cancer.[7] The ability to predict how a patient will respond to a particular treatment regimen is one of the most ambitious goals of personalized oncology. Also, pharmacogenomic-based studies of anti-cancer agents are complicated by the existence of additional somatic mutations within the tumor, as well as by the genetic heterogeneity of the population. One of the most important issues to address in pharmacogenomics is sample size, taking into account both the time required to complete the study population and the costs implied. While certain tools have been designed and approved for genetic testing in lung cancer, Lungscape and other tools for testing in early lung cancer are not readily available. Recently, the Pervenio™ Lung RS (risk score) (Life Technologies/Thermo Fisher Scientific)—a 14-gene expression assay using quantitative PCR—became available. Pervenio™ is based on two large independent studies of 1800 patients. In the validation cohort, 5-year overall survival was 49.2% (42.2–55.8) in high-risk patients compared to 58.3% (48.9–66.6) and 71.4% for the intermediate and low-risk groups, respectively.[8] This platform is intended to improve risk stratification of patients with early-stage, non-squamous NSCLC at high risk of mortality after surgical treatment.

2.1.2 The Molecular Diagnostics Approach

Individual patients' tumor biology and tumor heterogeneity are increasingly becoming important factors to consider when selecting cancer treatment. Similarly, there is now more emphasis on understanding the mechanisms of carcinogenesis and how these can be exploited when designing new therapeutic agents. Tumorigenesis in humans is a multistep process involving genetic alterations that drive the progressive transformation of normal cells to malignant types. Deregulated processes involved in tumorigenesis, such as regulation of cell cycle progression, angiogenesis, and apoptosis provide rational targets for novel therapies.[9]

Several technologies have been developed to detect cancer without the need for biopsy or surgical procedure; these new techniques also allow tumor cells to be studied at the molecular level.[10] Some of the most commonly used and useful assays are analysis of circulating tumor cells, mutation-specific PCR in circulating DNA, proteomic approaches to study serum or plasma, molecular-level images of tumors in situ, and assessment of autoantibodies specific for tumor cells.[11]

The majority of the molecular alterations we know about today were brought to light using technologies that detect changes in DNA content or sequence, transcription from DNA to messenger RNA or microRNA, production of proteins, or synthesis of various metabolic products. Table 2.1 summarizes the most common technologies and information obtained from each of them.[10,11]

In the past decade, there have been considerable improvements in the way human tumors are characterized. Knowledge of cancer at the molecular level has therefore increased greatly, and this has catalyzed a shift towards targeted therapies for cancer. However, there has been much less progress in the development of clinical

Table 2.1 Commonly used techniques in cancer research

Technique	Information obtained
DNA copy-number assessment	Comparative genome hybridization to DNA microarrays
Mutation screening	DNA sequencing Mass-spectrometry-based genotyping Mutation-specific PCR
Gene-expression profiling	DNA microarrays Multiplex PCR
MicroRNA-expression profiling	DNA microarrays Multiplex PCR
Proteomic profiling	Mass spectrometry
Phosphoproteomic profiling	Mass spectrometry after immunoprecipitation with phosphotyrosine-specific antibodies
Metabolomic profiling	Mass spectrometry

tools to determine which patients are most likely to benefit from particular targeted therapies or what the optimum dosage could be.

Biomarkers offer an innovative opportunity to obtain more information about each patient[4] and can be classified as prognostic, predictive, or pharmacodynamic biomarkers, depending on their potential use (Fig. 2.2). Prognostic biomarkers provide information

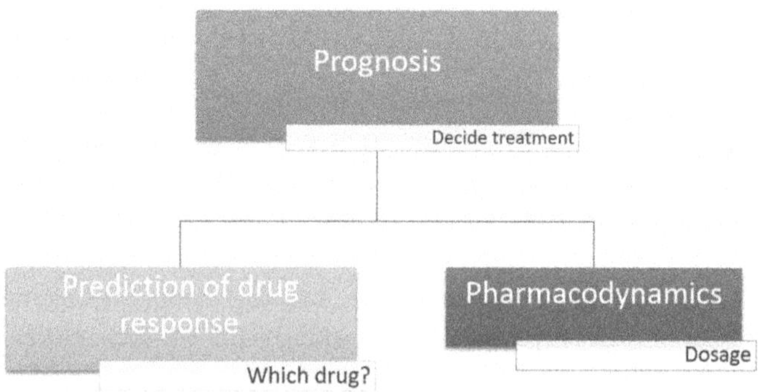

Figure 2.2 Types of biomarkers.

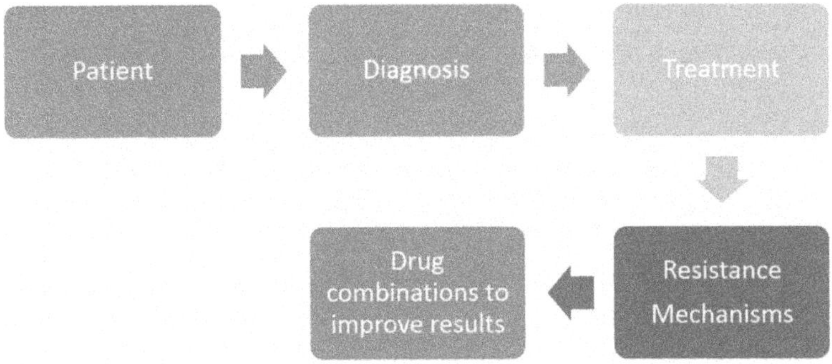

Figure 2.3 Envisaging the future of personalized precision medicine for cancer treatment.

about possible patient outcome and guide treatment selection.[12] Predictive biomarkers inform on the probability of positive or negative response to a specific treatment. Finally, pharmacodynamics biomarkers help the specialists to decide the optimum dose.[13] It is therefore possible to use biomarkers to help choose the most beneficial treatment and most suitable dose for each patient.

The solution to the inevitable challenge of polygenic cancer drug-resistance is to identify not only all exploitable molecular abnormalities but also the full range of resistance mechanisms and thereafter to employ precision combinatorial targeted therapy strategies matched specifically to the fully defined tumor profile (shown in Fig. 2.3).[14,15]

2.1.3 Conventional Chemotherapy

2.1.3.1 Cisplatin

Cisplatin, a platinum-based chemotherapeutic drug, has been used for over 30 years in a wide variety of cancers with varying degrees of success. It was the first member of a class of platinum-containing anti-cancer drugs, which now also includes carboplatin and oxaliplatin. These platinum complexes react in vivo, binding to and causing crosslinking of DNA, which ultimately triggers apoptosis (programmed cell death); cells able to remove these adducts are resistant to the compound. Many proteins involved in DNA damage-response (DDR) machinery have a role in repairing cross-links by platinum.[16] Specifically, cisplatin has been used to treat late stage NSCLC as the standard of care. However, therapeutic outcomes vary from patient to patient. Considerable efforts have been made to identify biomarkers that can be used to predict cisplatin sensitivity.[17]

Excision repair cross-complementing group 1 (ERCC1) has a crucial role in nucleotide excision repair (NER) pathway, which is one of DDR machinery. So far, many studies have suggested that ERCC1 at the level of protein, messenger RNA, or germline DNA could be a prognostic or predictive biomarker in NSCLC patients treated with platinum doublets, though contradictory results have also been reported.[18] A recent report in *The New England Journal of Medicine* by Friboulet and colleagues addressed the elegant validation of ERCC1 protein expression as a biomarker for adjuvant platinum-based chemotherapy and provided insights into the methodology to assess it.[19] Further trials have been performed to validate the predictive or prognostic effect of ERCC1.[20] Nevertheless, in these studies the discordance of ERCC1 H score between old and new batches of the 8F1 antibody and the point that four ERCC1 protein isoforms were heterogeneously expressed by alternative splicing in tumor samples with different functions in repairing platinum-DNA adduct, meant that no conclusive results could be extracted. Therefore the development of a specific antibody for functional isoform (ERCC1-202) will be necessary to more accurately predict the benefit from platinum.[18]

2.1.3.2 Pemetrexed

Pemetrexed (Alimta®, Ely Lilly) is chemically similar to folic acid and belongs to the class of chemotherapy drugs known as folate antimetabolites. It works by inhibiting three enzymes used in purine and pyrimidine synthesis—thymidylate synthase (TS), dihydrofolatereductase (DHFR), and glycinamideribonucleotideformyl transferase (GARFT).[21] By inhibiting the formation of precursor purine and pyrimidine nucleotides, pemetrexed prevents the formation of DNA and RNA required for growth and survival of both normal and cancer cells. In 2008, a phase III study by Scagliotti et al. comparing cisplatin/pemetrexed vs. cisplatin/gemcitabine showed for the first time a survival difference in favor of cisplatin/pemetrexed in two histologic groups (adenocarcinoma and large cell carcinoma). The first group had significantly better survival (12.6 vs. 10.9 months),[22] something which may be explained by the fact that thymidylate synthase levels are generally lower in adenocarcinoma. Pre-clinical data suggests that overexpression of TS correlates with reduced sensitivity to pemetrexed;[23] this is corroborated by a recent meta-analysis. The authors concluded that TS may be a suitable marker of sensitivity to pemetrexed-based chemotherapy in NSCLC patients.[24] Nevertheless, this is not yet a standard recommendation for clinical practice.

2.1.3.3 Gemcitabine

This chemotherapeutic agent is one of the most widely used pyrimidine analogues, with a very well established role in first-line treatment of advanced NSCLC. Several studies have examined molecular determinants of sensitivity to gemcitabine. Potential candidates to predict response include genes encoding drug metabolism enzymes, transport across membranes, or target proteins. One example is human equilibrative nucleoside transporter 1 (hENT1), which transports gemcitabine into cells. hENT1 basal expression levels have been correlated with IC50 values for gemcitabine in several NSCLC cell lines.[25] Also, expression of hENT1 has been analyzed by IHC, suggesting that absence of hENT1 expression may be used to predict response to gemcitabine-based chemotherapy.[26] Nevertheless, prospective studies are still required to validate this and the most accurate determination method remains to be decided.

Personalized treatment of lung cancer, although quite common in daily clinical practice in some parts of the world, is not widespread in others. There is an urgent need to improve treatment strategies, especially for those patients with advanced disease who are not candidates for targeted treatments such as erlotinib for EGFR mutated patients or crizotinib for ALK translocated patients. The Spanish Lung Cancer Group (SLCG) recently presented the results of the BRCA1-RAP80 Expression Customization (BREC) trial, a good example of the complexity of modern pharmacogenomics. Investigators allocated patients to three different customized therapy arms according to RAP80 and BRCA1 expression and results of the interim analysis were presented at the 15th World Congress on Lung Cancer (WCLC) in Sydney. The study was closed early due to negative results, possibly due to the poor capacity of RAP80 as a predictor.[27] As commented, this illustrates the difficulty of finding accurate biomarkers.

2.1.3.4 Taxanes

Taxanes agents are diterpenes produced by the plants of the genus Taxus (yews), and include paclitaxel (Taxol) and docetaxel (Taxotere). Taxanes present difficulties in formulation as medicines because they are poorly soluble in water due to their lack of acid and basic groups. Their action is based on disruption of microtubule function; microtubules are essential to cell division, and taxanes stabilize guanosinediphosphate (GDP)-bound tubulin in the microtubule, thereby inhibiting the process of cell division. Thus, in essence, taxanes are mitotic inhibitors. Taxanes are well established as potent chemotherapy agents for metastatic breast cancer (MBC) and early breast cancer (EBC). However, their therapeutic usefulness is limited by de novo refractoriness or acquired resistance, common drawbacks to most anti-cancer cytotoxics.[28]

Thymidine phosphorylase (TP) allows classification of patients as high-expression and low-expression according to the median value of its expression, with significantly longer time to progression (TTP) in the high-TP group. In some trials, it has been observed that TP expression may be a prognostic factor in breast cancer patients treated with capecitabine-based first-line chemotherapy and βIII-tubulin can predict outcome of capecitabine in combination with taxanes as first-line chemotherapy.[29]

Over-expression of the MDR-1 gene product Pgp has been extensively studied in vitro in association with taxane resistance, but data are conflicting. Similarly, the target components microtubules failed to confirm such associations. Moreover, little consensus has been reached for reported associations between taxane-sensitivity and mutated p53, or taxane-resistance and overexpression of Bcl-2, Bcl-xL, or NFkB. In contrast, the most notable finding is that pharmaceutical down-regulation of HER-2 appears to reverse taxane resistance.[29]

2.2 Genetic Alterations and New Potential Targets

In NSCLC, numerous genes involved in tumor proliferation are the target of agents currently in various stages of clinical development: EGFR, HER2 (human epidermal growth factor receptor 2), ROS1 (reactive oxygen species 1), BRAF (v-raf murine sarcoma viral oncogene homologue B1), MAPK (mitogen-activated protein kinase), c-MET (c-mesenchymal-epithelial transition), PTEN (phosphatase and tensin homolog), FGFR (fibroblast growth factor receptor), DDR2 (discoidin domain receptor 2), PIK3CA (phosphatidylinositol-4,5-bisphosphate3-kinase, catalytic subunit alpha), RET (rearranged during transfection), AKT (protein kinase B) and ALK (anaplastic lymphoma kinase), among others. The activity of these oncogenic targets occurs through various pathways such as DRC-signal transduction, phosphoinositide 3-kinase-AKT-mTOR, RAS-RAF-MEK, etc. To date, there are five approved targeted therapies for treatment of advanced or metastatic NSCLC: gefitinib, erlotinib, and afatinib for EGFR mutated patients, crizotinib for ALK translocated patients and bevacizumab which currently lacks a reliable pre-treatment biomarker. Moreover, oncogenic mutant proteins are subject to regulation by protein trafficking pathways, specifically through the heat shock protein 90 system. Drug combinations affecting various nodes in these signaling and intracellular processes have been demonstrated to be synergistic and advantageous in overcoming treatment resistance compared with monotherapy approaches. Understanding the role of the tumor microenvironment in development and maintenance of the malignant phenotype has also provided additional therapeutic approaches. More recently, improved knowledge of tumor immunology has set the stage for promising immunotherapies in NSCLC. The main molecular alterations are listed in Table 2.2.

Table 2.2 Molecular alterations in NSCLC

Gene	Frequency (%) ADC	Frequency (%) SCC	Available therapies	Therapies under study
EGFR[63]	10	2–3	Erlotinib, gefitinib, afatinib	AZD9291, CO-186, HM61713
ALK[64]	3–5	<1	Crizotinib, ceritinib	AP26113, alectinib, ganetespib, PF-06463922
MET[65]	2–4	NA	Crizotinib	Tivantinib, cabozantinib, INC280, onartuzumab
ROS1[66]	1–2	NA	Crizotinib	PF-06463922
FGFR1[67]	NA	20	NA	Dovitinib, ponatinib, AZD4547, BGJ398
FGFR2[68]	3	3	NA	Dovitinib, ponatinib, AZD4547, BGJ398
NTRK1[69]	1–2	NA	N/A	Crizotinib, lestaurtinib
RET[70]	1	N/A	N/A	Carbozantinib, vandetanib
HER2[71]	2–4	N/A	N/A	Neratinib, afatinib, lapatinib, trastuzumab
DDR2[72]	N/A	2–3	N/A	Dasatinib
BRAF[73]	1–6	4–5	N/A	Vemurafenib, dabrafenib, trametinib
KRAS[74]	15–25	1–2	N/A	Selumetinib + docetaxel

2.2.1 Receptor Tyrosine Kinases

2.2.1.1 EGFR inhibitors (first and second generation)

In 2000, results from preclinical trials suggested that EGFR inhibitors contrasted proliferation and angiogenesis, promoted or induced apoptosis and may have a synergistic effect with other standard cytotoxic cancer therapies.[30]

Gefitinib [4-(3-chloro-4-fluroanilino)-7-methoxy-6-(3-morpholinopropoxy) quinazoline] is a low molecular weight synthetic aniline-quinazoline oral selective reversible inhibitor of EGFR tyrosine kinase (TK). Erlotinib [6,7-bis(2-methoxy-ethoxy)-quinazolin-4-yl-(3-ethynulphenyl)amine] is a quinazoline derivative which reversibly inhibits the kinase activity of purified EGFR and autophosphorylation in intact cells in vitro.[31] EGFR tyrosine kinases inhibitors (TKIs) target cytoplasmic-domain phosphorylation,[32] which in turn results in apoptosis and growth arrest of EGFR-dependent cells by inhibition of signaling pathways such as PI3K (phosphatidylinositol-3-kinase)-AKT, Jak (Janus kinase)-STAT (signal transducer and activator of transcription) and MAPK (mitogen-activated protein kinase).[33]

Second-generation EGFR inhibitors (irreversible inhibitors) such as afatinib, dacomitinib, and neratinib are pan-ErbB inhibitors that form covalent bonds to the receptor, and therefore are biologically active even in the presence of the T790M mutation.[34]

2.2.1.2 ALK rearrangement (first and second generation)

Some types of cancer are sensitive to specific inhibitors targeting a mutated pathway due to specific genetic lesions that work as proliferation drivers in these cells. Activating mutations or translocations of the anaplastic lymphoma kinase (ALK) gene have been identified in several types of cancer, including NSCLC. The product of this translocation (EML4-ALK) is a fusion gene that encodes a cytoplasmic chimeric protein with constitutive kinase activity. Therefore, the hypothesis is that patient outcomes may be optimized by testing tumors for specific mutated pathways and directing therapies against those mutant pathways. This methodology led to one of the fastest FDA approvals in history: crizotinib was granted accelerated FDA approval less than five years after the first ALK positive NSCLC patient was included in crizotinib trials. The accelerated approval was based on two single-arm trials, one with 136 and the other with 119 locally advanced or metastatic ALK-positive NSCLC patients. The primary endpoint in both trials was objective response rate (ORR). In the first study this was 50%, while in the second it was 61%. These results included complete responses in 1% of treated patients, irrespective of their performance status or number of prior chemotherapy regimens.[35] Final approval in 2013 was based on the demonstration of better progression-free survival (PFS) and ORR with crizotinib compared to chemotherapy in NSCLC patients harboring ALK rearrangements with disease progression after standard chemotherapy treatment.[36] Nevertheless, despite impressive results with crizotinib, we now know that most patients will eventually relapse and develop resistance to targeted therapies.

Second-generation ALK inhibitors such as ceritinib (LDK378) and alectinib (AP26113) have been evaluated in clinical trials. Response rates of up to 60% and predicted disease control rates of approximately 90% have been reported in crizotinib-resistant ALK positive NSCLC patients.[37]

LDK378 (ceritinib)—an oral ATP-competitive ALK TKI—achieved a 56% response rate in patients previously treated with crizotinib, both in those with the presence of some ALK-resistance mutations and those without. As expected, accelerated FDA approval was granted to ceritinib in April 2014 for ALK positive patients with metastatic NSCLC intolerant to crizotinib or with progressive disease.

Early clinical results of ganetespib (a HSP90 inhibitor) have demonstrated that it may constitute a different strategy to target ALK by inducing antitumor responses as well as overcoming acquired resistance.[38] The combination of ganetespib with crizotinib showed superior antitumor efficacy compared with monotherapy in in vivo experiments.[39]

2.2.1.3 ROS1

One to two percent of NSCLC patients have a chromosomal rearrangement involving the ROS1 gene (c-ros oncogene 1).[40] The ROS1 (chromosome 6q22) encodes a receptor tyrosine kinase of the insulin receptor family, which has downstream signaling via the MAPK pathway through RAS phosphorylation.[41] ROS1 fusion partners in NSCLC include FIG, CD74, SLC34A2, and SDC4, which lead to oncogenic transformation and constitutive kinase activity in cell culture and/or in vivo.[40,42]

According to preclinical data, ROS1 can be targeted by EML4-ALK inhibitors, thus suggesting an off-target effect.[43] So far, no specific ligand has been identified for the ROS1 tyrosine kinase. However, clinical findings of a cohort of patients with the ROS1 rearrangement showed that they share features with EML4-ALK-translocated patients, thus leading to the use of crizotinib as a targeted treatment option with high response rates. Nevertheless, little is known regarding the prognostic and predictive value, as well as the clinical presentation of patients harboring a ROS1 rearrangement.

2.2.2 Epigenetic Factors

Epigenetics is the study of cellular and physiological traits not caused by changes in the DNA sequence, i.e., relevant changes to the genome that do not involve a change in the nucleotide sequence.[44] Mechanisms of epigenetic silencing of tumor suppressor genes and activation of oncogenes include alteration in CpG island methylation patterns, histone modifications, and dysregulation of DNA proteins. Understanding epigenetic mechanisms holds great promise for cancer prevention, detection, and therapy.

DNA methylation is an important regulator in genes transcription as it is related to non-programmed silencing genes.[45] The most common mechanisms of DNA methylation are hypermethylation and hypomethylation. Hypermethylation occurs when the CpG regions have a high amount of 5-methylcitosine, this produces gene silencing which is strongly involved in certain cancer processes, while hypomethylation has been related to cancer through different mechanisms.[46] CpG regions are areas with high amounts of cytosine and guanines bases bonded by phosphate groups. CpG sites located on encoded regions have unmethylated cytokines and guanine bases and the genes are expressed. However, when these bases are methylated the corresponding genes are silenced (Fig. 2.4). Therefore when these genes act as tumor suppressors and are methylated, development of a tumoral mass is favored.

Histones are highly alkaline proteins found in eukaryotic cell nuclei that package and order the DNA into structural units called nucleosomes located in the nucleus. Within the nucleus DNA is packaged as chromatin in the nucleosome and composed of an octamer of four different histones (H3, H4, H2A, H2B). Therefore histones are the chief protein components of chromatin, acting as spools around which DNA winds, and playing a role in gene regulation.

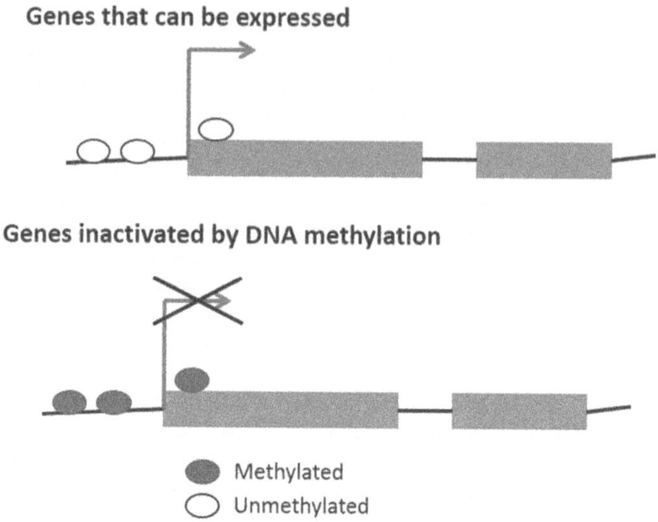

Figure 2.4 DNA methylation.

Histone modifications are reversible modifications in the amino-terminal tail of histones which display a large number of modified residues. Acetylation, methylation and phosphorylation are examples of histone modifications (Fig. 2.5). These modifications are catalyzed by distinct enzymes and divide the genome into "active"

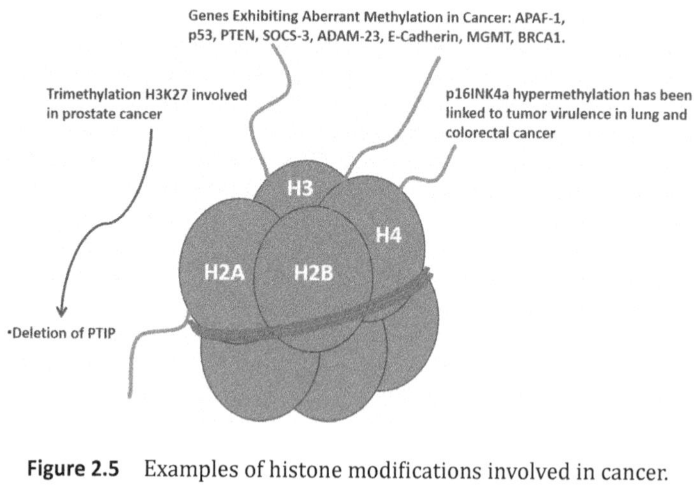

Figure 2.5 Examples of histone modifications involved in cancer.

or euchromatin, in which DNA is accessible for transcription, and "inactive" or heterochromatin, in which DNA is inaccessible for transcription due to misregulation of some chromatin functions. As an example, prostate cancer has been associated with gene silencing by CpG island hypermethylation. GSTP1, APC, RASSF1a, PTGS2, and MDRI genes have been found to defend prostate cells against genomic damage caused by different oxidants or carcinogens.[47] This suggests that silencing of these genes will permit genetic damage to the prostate by oxidants and carcinogens.

Some epigenetic therapies in cancer include the DNMT inhibitors 5-azacytidine and 5-aza-20-deoxycytidine, which have been approved by the FDA for the treatment of various forms of cancer. These drugs have been shown to reactivate the cellular antitumor systems repressed by cancer, enabling the body to weaken the tumor.[48] Because of their wide ranging effects throughout the entire organism, all of these drugs have major side effects but do significantly increase survival rates. Figure 2.6 shows some of the repair processes following DNA damage.

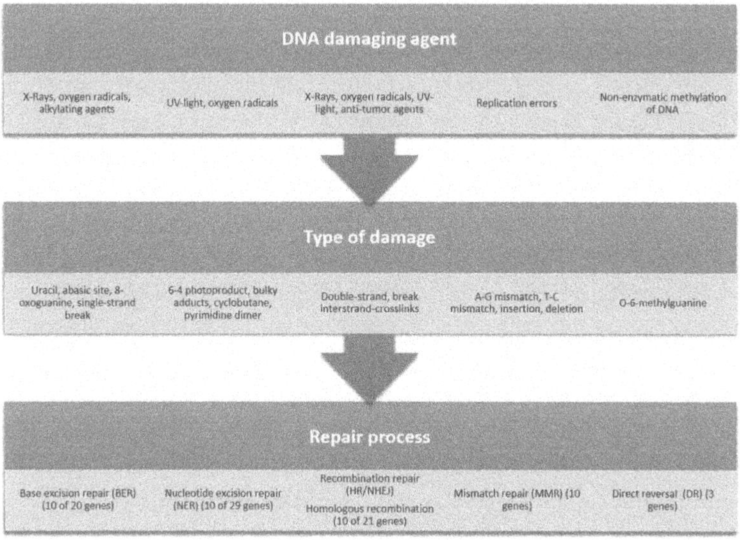

Figure 2.6 DNA damage, DNA repair and epigenetic-repair alterations in cancer.

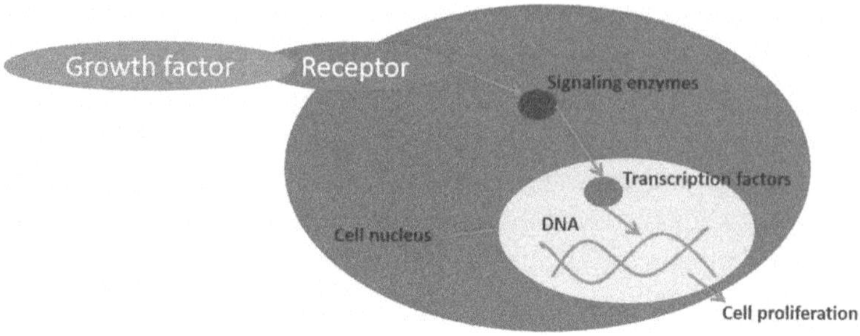

Figure 2.7 Normal growth pathway.

2.2.3 Transcription Factors

A transcription factor is a protein that is involved in the regulation of DNA transcription process but is not part of the RNA polymerase. Activity of these factors can occur through recognition or binding to specific DNA sequences, binding to other factors or direct binding to the RNA polymerase. Transcription factors are stimulated by cytoplasmic signals and are then able to regulate gene expression in the cellular nucleus and activate or suppress gene transcription (Fig. 2.7).

Some oncogenes that cause neoplastic lesions are the result of mutations in transcription factors. These mutations produce constant activity of transcription factors without the external signal requirement so regulation of the cell cycle is out of control. Some oncogenic transcription factors are Myc, STAT, Max, Myb, Fos, Jun, Rel. Traditionally, transcription factors were generally considered too difficult to target, and kinase pathways or cell surface proteins have long been popular alternative therapeutic targets. However, transcription factors are downstream effectors of many pathways and this, coupled with technological advances, has made them an attractive and realistic drug target.[49] Some of the components that could be targeted by drugs include cyclin-dependent kinases (CDKs), RNA polymerases (RNPI and RNPII), or components of associated transcriptional complexes.[50] These components are related to RNP enzymes and, although they are the expected targets for transcription inhibition, only a few drugs that directly affect them have been described, such as α-amanitin cyclic octapeptide and TAS-106, a cytidine analog. Their mechanisms of action involve binding to RNP to prevent DNA and RNA translocation in the α-amanitin case, while TAS-106 is able to induce apoptosis as well as decrease the transcription of several factors required for survival. However, due to its high hepatotoxicity, α-amanitin is not suitable for cancer treatment.[51]

Many drugs target CDKs as these are deregulated in cancer cells. Their inhibitors compete with ATP for the enzyme active site. Therefore, CDK inhibition results in RNPII hypophosphorylation.[51] The most commonly targeted CDKs are CDK7, CDK8 and CDK9. CDK7 is a component of basal transcription factor TFIIH that phosphorylates Serine 5 and 7 in the C-terminal domain (CTD) of the RNPII, which is important for promoter escape and recruitment of mRNA processing machinery during transcription.[52] CDK9 is also a component of P-TEFb, which, similar to CDK7, phosphorylates CTD of RNPII at serine 2 for transcription elongation.[53,54] The same activity is observed with the CDK8 kinase, which phosphorylates CTD of RNPII, resulting in inhibition of transcription initiation complex.

Finally, transcription can be disrupted via targeting of associated transcriptional complex components. In some cases, agents bind covalently to TFIIH and inhibit its ATPase activity. This action disrupts the opening of double-stranded DNA for RNPII transcription and repair as well as RNPI transcription. Therefore, cytotoxicity of these drugs is associated with transcriptional inhibition of anti-apoptotic factors and induction of apoptotic factors.[55,56] BRD3 and BRD4 are important emerging targets for treatment of various cancers. Displacement of BET bromodomains from chromatin prevents BRD3 and BRD4 reader activity. Inhibition of these factors has a generalized effect on RNPII transcription which causes downregulation of BCL-2, MYC and CDK6, thereby inducing cell cycle arrest and apoptosis.[57–59] Usually, transcription inhibition induces apoptosis by four possible mechanisms: altering the balance of apoptotic and anti-apoptotic factors to favor apoptosis, activating p53 and promoting its translocation to mitochondria, inhibiting DNA replication, and promoting accumulation of aberrant proteins in the nucleus. Moreover, oncogenes that are frequently overexpressed in cancer cells can be suppressed via transcription inhibition without affecting other genes (Fig. 2.8).

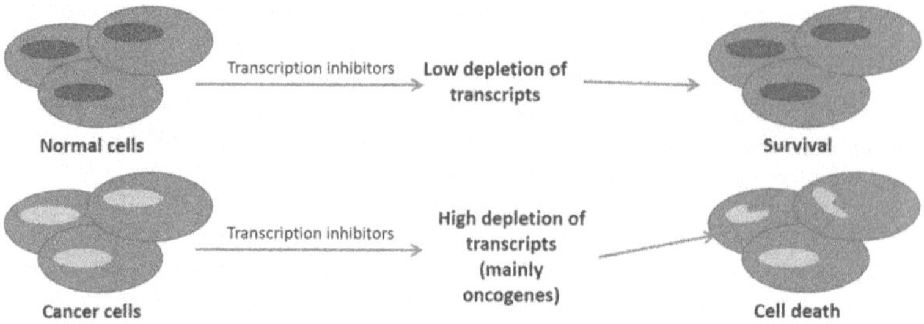

Figure 2.8 Transcription inhibitors affect differentially regulatory sequences of genes and trigger cell death in cancer cells.

2.2.4 Repurposing Drugs

Despite great advances to date, failure rates are progressively increasing. Cancer treatments are extremely costly with a lengthy design and testing process with poor safety and bioavailability and often limited efficacy. It is estimated that only one of every 5000 to 10,000 new anticancer agents will receive FDA approval, and a mere 5% of all oncology drugs in phase I trials will be approved. This has highlighted the need for alternative efforts in the development of cancer treatments.[60] Approximately 13 years of research and nearly 1.8 US billion dollars are required to take a new drug all the way from the bench to bed.[61]

We now know that almost every drug has more than one target and therefore can produce off-target side effects. If this interaction with the off-target is potent enough, the same drug could be used to treat more than one disease. Evaluation of established non-cancer drugs approved for other medical conditions but with known cancer targets and reliable biomarkers is known as "drug repurposing." This approach is predicted to soon become part of the tailored medicine approach and should be considered part of modern pharmacogenomics.[62] Drug repurposing of non-cancer drugs for their potential anticancer activities can provide the opportunity to rapidly advance therapeutic strategies into clinical trials. If this approach is followed, it will not be long before existing drugs known for their analgesic, antidiabetic, or immunosuppressant activities become part of the daily practice in oncology.

2.3 Conclusions

Fortunately, discovery of molecular mechanisms underlying lung cancer and development of possible targeted therapies have accelerated in recent years. The knowledge generated has helped to improve both the length and quality of life in NSCLC patients. It is to be hoped that it will not be long before there is a specific treatment available for every patient.

References

1. Siegel RL, Miller KD, Jemal A. Cancer statistics, 2015. *CA Cancer J Clin.* 2015;65(1):5–29.
2. Scagliotti G, Novello S. Adjuvant chemotherapy after complete resection for early stage NSCLC. *Lung Cancer.* 2003;42(2):47–51.
3. Kobayashi N, Toyooka S, Soh J, et al. Risk factors for recurrence and unfavorable prognosis in patients with stage I non-small cell lung cancer and a tumor diameter of 20 mm or less. *J Thorac Oncol.* 2007;2(9):808–812.
4. Hsu C-P, Hsia J-Y, Chang G-C, et al. Surgical-pathologic factors affect long-term outcomes in stage IB (pT2 N0 M0) non-small cell lung cancer: a heterogeneous disease. *J Thorac Cardiovasc Surg.* 2009;138(2):426–433.
5. Van de Vijver MJ, He YD, van't Veer LJ, et al. A gene-expression signature as a predictor of survival in breast cancer. *N Engl J Med.* 2002;347(25):1999–2009.
6. Eschrich S, Yang I, Bloom G, et al. Molecular staging for survival prediction of colorectal cancer patients. *J Clin Oncol.* 2005;23(15):3526–3535.
7. Hoshida Y, Villanueva A, Kobayashi M, et al. Gene expression in fixed tissues and outcome in hepatocellular carcinoma. *N Engl J Med.* 2008;359(19):1995–2004.
8. Kratz JR, He J, Van Den Eeden SK, et al. A practical molecular assay to predict survival in resected non-squamous, non-small-cell lung cancer: development and international validation studies. *Lancet.* 2012;379(9818):823–832.
9. Yarden Y, Baselga J, Miles D. Molecular approach to breast cancer treatment. *Semin Oncol.* 2004;31(5 Suppl 10):6–13.

10. Sawyers CL. The cancer biomarker problem. *Nature*. 2008;452(7187): 548–552.
11. Hunter DJ, Khoury MJ, Drazen JM. Letting the genome out of the bottle–will we get our wish? *N Engl J Med*. 2008;358(2):105–107.
12. Ratain MJ, Glassman RH. Biomarkers in phase I oncology trials: signal, noise, or expensive distraction? *Clin Cancer Res*. 2007;13(22 Pt 1):6545–6548.
13. Sharma S V, Bell DW, Settleman J, Haber DA. Epidermal growth factor receptor mutations in lung cancer. *Nat Rev Cancer*. 2007;7(3):169–181.
14. Gonzalez de Castro D, Clarke PA, Al-Lazikani B, Workman P. Personalized cancer medicine: molecular diagnostics, predictive biomarkers, and drug resistance. *Clin Pharmacol Ther*. 2013;93(3):252–259.
15. Al-Lazikani B, Banerji U, Workman P. Combinatorial drug therapy for cancer in the post-genomic era. *Nat Biotechnol*. 2012;30(7):679–692.
16. Ciccia A, Elledge SJ. The DNA damage response: making it safe to play with knives. *Mol Cell*. 2010;40(2):179–204.
17. Rose MC, Kostyanovskaya E, Huang RS. Pharmacogenomics of cisplatin sensitivity in non-small cell lung cancer. *Genomics Proteomics Bioinformatics*. 2014;12(5):198–209.
18. Ryu JS, Memon A, Lee S-K. ERCC1 and personalized medicine in lung cancer. *Ann Transl Med*. 2014;2(4):32.
19. Friboulet L, Olaussen KA, Pignon J-P, et al. ERCC1 isoform expression and DNA repair in non-small-cell lung cancer. *N Engl J Med*. 2013;368(12):1101–1110.
20. Winton T, Livingston R, Johnson D, et al. Vinorelbine plus cisplatin vs. observation in resected non-small-cell lung cancer. *N Engl J Med*. 2005;352(25):2589–2597.
21. McLeod HL, Cassidy J, Powrie RH, et al. Pharmacokinetic and pharmacodynamic evaluation of the glycinamide ribonucleotide formyltransferase inhibitor AG2034. *Clin Cancer Res*. 2000;6(7):2677–2684.
22. Scagliotti GV, Parikh P, von Pawel J, et al. Phase III study comparing cisplatin plus gemcitabine with cisplatin plus pemetrexed in chemotherapy-naive patients with advanced-stage non-small-cell lung cancer. *J Clin Oncol*. 2008;26(21):3543–3551.
23. Sigmond J, Backus HHJ, Wouters D, Temmink OH, Jansen G, Peters GJ. Induction of resistance to the multitargeted antifolate Pemetrexed (AL-

IMTA) in WiDr human colon cancer cells is associated with thymidylate synthase overexpression. *Biochem Pharmacol.* 2003;66(3):431–438.

24. Wang L, Wang R, Pan Y, Sun Y, Zhang J, Chen H. The pemetrexed-containing treatments in the non-small cell lung cancer is -/low thymidylate synthase expression better than +/high thymidylate synthase expression: a meta-analysis. *BMC Cancer.* 2014;14:205.
25. Achiwa H, Oguri T, Sato S, Maeda H, Niimi T, Ueda R. Determinants of sensitivity and resistance to gemcitabine: the roles of human equilibrative nucleoside transporter 1 and deoxycytidine kinase in non-small cell lung cancer. *Cancer Sci.* 2004;95(9):753–757.
26. Oguri T, Achiwa H, Muramatsu H, et al. The absence of human equilibrative nucleoside transporter 1 expression predicts nonresponse to gemcitabine-containing chemotherapy in non-small cell lung cancer. *Cancer Lett.* 2007;256(1):112–119.
27. Moran T, Cobo M, Domine M, et al. Interim analysis of the Spanish Lung Cancer Group (SLCG) BRCA1-RAP80 Expression Customization (BREC) randomized phase III trial of customized therapy in advanced non-small cell lung cancer (NSCLC) patients (p) (NCT00617656/GECP-BREC). *ASCO Meet Abstr.* 2013;31(18 Suppl):LBA8002.
28. Murray S, Briasoulis E, Linardou H, Bafaloukos D, Papadimitriou C. Taxane resistance in breast cancer: mechanisms, predictive biomarkers and circumvention strategies. *Cancer Treat Rev.* 2012;38(7):890–903.
29. Zhao H-Y, Huang H, Hu Z-H, et al. Evaluations of biomarkers associated with sensitivity to 5-fluorouracil and taxanes for recurrent/advanced breast cancer patients treated with capecitabine-based first-line chemotherapy. *Anticancer Drugs.* 2012;23(5):534–542.
30. Ciardiello F, Caputo R, Bianco R, et al. Antitumor effect and potentiation of cytotoxic drugs activity in human cancer cells by ZD-1839 (Iressa), an epidermal growth factor receptor-selective tyrosine kinase inhibitor. *Clin Cancer Res.* 2000;6(5):2053–2063.
31. Ciardiello F, Tortora G. A novel approach in the treatment of cancer: targeting the epidermal growth factor receptor. *Clin Cancer Res.* 2001;7(10):2958–2970.
32. Dancey J, Sausville EA. Issues and progress with protein kinase inhibitors for cancer treatment. *Nat Rev Drug Discov.* 2003;2(4):296–313.
33. Mendelsohn J, Baselga J. Status of epidermal growth factor receptor antagonists in the biology and treatment of cancer. *J Clin Oncol.* 2003;21(14):2787–2799.

34. Ninomiya T, Takigawa N, Ichihara E, et al. Afatinib prolongs survival compared with gefitinib in an epidermal growth factor receptor-driven lung cancer model. *Mol Cancer Ther.* 2013;12(5):589–597.
35. Camidge DR, Bang Y-J, Kwak EL, et al. Activity and safety of crizotinib in patients with ALK-positive non-small-cell lung cancer: updated results from a phase 1 study. *Lancet Oncol.* 2012;13(10):1011–1019.
36. Shaw AT, Kim D-W, Nakagawa K, et al. Crizotinib versus chemotherapy in advanced ALK-positive lung cancer. *N Engl J Med.* 2013;368(25):2385–2394.
37. Doebele RC. A nice problem to have: when ALK inhibitor therapy works better than expected. *J Thorac Oncol.* 2014;9(4):433–435.
38. Sang J, Acquaviva J, Friedland JC, et al. Targeted inhibition of the molecular chaperone Hsp90 overcomes ALK inhibitor resistance in non-small cell lung cancer. *Cancer Discov.* 2013;3(4):430–443.
39. Richards MW, Law EWP, Rennalls LP, et al. Crystal structure of EML1 reveals the basis for Hsp90 dependence of oncogenic EML4-ALK by disruption of an atypical β-propeller domain. *Proc Natl Acad Sci U S A.* 2014;111(14):5195–5200.
40. Takeuchi K, Soda M, Togashi Y, et al. RET, ROS1 and ALK fusions in lung cancer. *Nat Med.* 2012;18(3):378–381.
41. Oxnard GR, Binder A, Jänne PA. New targetable oncogenes in non-small-cell lung cancer. *J Clin Oncol.* 2013;31(8):1097–1104.
42. Rimkunas VM, Crosby KE, Li D, et al. Analysis of receptor tyrosine kinase ROS1-positive tumors in non-small cell lung cancer: identification of a FIG-ROS1 fusion. *Clin Cancer Res.* 2012;18(16):4449–4457.
43. McDermott U, Iafrate AJ, Gray NS, et al. Genomic alterations of anaplastic lymphoma kinase may sensitize tumors to anaplastic lymphoma kinase inhibitors. *Cancer Res.* 2008;68(9):3389–3395.
44. Ledford H. Language: Disputed definitions. *Nature.* 2008;455(7216):1023–1028.
45. Jin B, Li Y, Robertson KD. DNA methylation: superior or subordinate in the epigenetic hierarchy? *Genes Cancer.* 2011;2(6):607–617.
46. Horvath S. DNA methylation age of human tissues and cell types. *Genome Biol.* 2013;14(10):R115.
47. Gurel B, Iwata T, Koh CM, Yegnasubramanian S, Nelson WG, De Marzo AM. Molecular alterations in prostate cancer as diagnostic, prognostic, and therapeutic targets. *Adv Anat Pathol.* 2008;15(6):319–331.

48. Kowluru RA, Santos JM, Mishra M. Epigenetic modifications and diabetic retinopathy. *Biomed Res Int.* 2013;2013:635284.

49. Redmond AM, Carroll JS. Defining and targeting transcription factors in cancer. *Genome Biol.* 2009;10(7):311.

50. Villicaña C, Cruz G, Zurita M. The basal transcription machinery as a target for cancer therapy. *Cancer Cell Int.* 2014;14(1):18.

51. Bensaude O. Inhibiting eukaryotic transcription. Which compound to choose? How to evaluate its activity? *Transcription.* 2014;2(3):103–108.

52. Akhtar MS, Heidemann M, Tietjen JR, et al. TFIIH kinase places bivalent marks on the carboxy-terminal domain of RNA polymerase II. *Mol Cell.* 2009;34(3):387–393.

53. Peng J, Liu M, Marion J, Zhu Y, Price DH. RNA polymerase II elongation control. *Cold Spring Harb Symp Quant Biol.* 1998;63:365–370.

54. Fu TJ, Peng J, Lee G, Price DH, Flores O. Cyclin K functions as a CDK9 regulatory subunit and participates in RNA polymerase II transcription. *J Biol Chem.* 1999;274(49):34527–34530.

55. Manzo SG, Zhou Z-L, Wang Y-Q, et al. Natural product triptolide mediates cancer cell death by triggering CDK7-dependent degradation of RNA polymerase II. *Cancer Res.* 2012;72(20):5363–5373.

56. Vispé S, DeVries L, Créancier L, et al. Triptolide is an inhibitor of RNA polymerase I and II-dependent transcription leading predominantly to down-regulation of short-lived mRNA. *Mol Cancer Ther.* 2009;8(10):2780–2790.

57. Luo Z, Lin C, Guest E, et al. The super elongation complex family of RNA polymerase II elongation factors: gene target specificity and transcriptional output. *Mol Cell Biol.* 2012;32(13):2608–2617.

58. Filippakopoulos P, Qi J, Picaud S, et al. Selective inhibition of BET bromodomains. *Nature.* 2010;468(7327):1067–1073.

59. Dawson MA, Prinjha RK, Dittmann A, et al. Inhibition of BET recruitment to chromatin as an effective treatment for MLL-fusion leukaemia. *Nature.* 2011;478(7370):529–533.

60. Zamboni WC, Torchilin V, Patri AK, et al. Best Practices in Cancer Nanotechnology: Perspective from NCI Nanotechnology Alliance. *Clin Cancer Res.* 2012;18(12):3229–3241.

61. Paul SM, Mytelka DS, Dunwiddie CT, et al. How to improve R&D productivity: the pharmaceutical industry's grand challenge. *Nat Rev Drug Discov.* 2010;9(3):203–214.

62. Gupta SC, Sung B, Prasad S, Webb LJ, Aggarwal BB. Cancer drug discovery by repurposing: teaching new tricks to old dogs. *Trends Pharmacol Sci.* 2013;34(9):508–517.
63. Paez JG, Jänne PA, Lee JC, et al. EGFR mutations in lung cancer: correlation with clinical response to gefitinib therapy. *Science.* 2004;304(5676):1497–1500.
64. Soda M, Choi YL, Enomoto M, et al. Identification of the transforming EML4-ALK fusion gene in non-small-cell lung cancer. *Nature.* 2007;448(7153):561–566.
65. Engelman JA, Zejnullahu K, Mitsudomi T, et al. MET amplification leads to gefitinib resistance in lung cancer by activating ERBB3 signaling. *Science.* 2007;316(5827):1039–1043.
66. Rikova K, Guo A, Zeng Q, et al. Global survey of phosphotyrosine signaling identifies oncogenic kinases in lung cancer. *Cell.* 2007;131(6):1190–1203.
67. Weiss J, Sos ML, Seidel D, et al. Frequent and focal FGFR1 amplification associates with therapeutically tractable FGFR1 dependency in squamous cell lung cancer. *Sci Transl Med.* 2010;2(62):62ra93.
68. Comprehensive genomic characterization of squamous cell lung cancers. *Nature.* 2012;489(7417):519–525.
69. Vaishnavi A, Capelletti M, Le AT, et al. Oncogenic and drug-sensitive NTRK1 rearrangements in lung cancer. *Nat Med.* 2013;19(11):1469–1472.
70. Kohno T, Ichikawa H, Totoki Y, et al. KIF5B-RET fusions in lung adenocarcinoma. *Nat Med.* 2012;18(3):375–377.
71. Stephens P, Hunter C, Bignell G, et al. Lung cancer: intragenic ERBB2 kinase mutations in tumours. *Nature.* 2004;431(7008):525–526.
72. Hammerman PS, Sos ML, Ramos AH, et al. Mutations in the DDR2 kinase gene identify a novel therapeutic target in squamous cell lung cancer. *Cancer Discov.* 2011;1(1):78–89.
73. Paik PK, Arcila ME, Fara M, et al. Clinical characteristics of patients with lung adenocarcinomas harboring BRAF mutations. *J Clin Oncol.* 2011;29(15):2046–2051.
74. Mascaux C, Iannino N, Martin B, et al. The role of RAS oncogene in survival of patients with lung cancer: a systematic review of the literature with meta-analysis. *Br J Cancer.* 2005;92(1):131–139.

Chapter 3

Genome-Based Personalized Medicine in Liver Cancer

Jae-Jun Shim and Ju-Seog Lee
Department of Systems Biology, Division of Cancer,
The University of Texas MD Anderson Cancer Center, 7435 Fannin Street,
Houston, TX 77054, USA
jlee@mdanderson.org

Our understanding of liver cancer has rapidly progressed through analysis of genomic/epigenomic/proteomic data from tumors. To take advantage of the advancement, it will be necessary to develop systematic approaches that can uncover potential biomarkers reflecting genetic alterations, prognosis, or response to treatments. In this chapter, updated results and future perspectives of liver cancer toward personalized medicine using genomics, bioinformatics, and systems biology approaches were summarized.

3.1 Introduction

3.1.1 Epidemiology of Liver Cancer

Liver cancer is one of major threat in public health. Estimated number of new cases and death from liver cancer was 748,300

Cancer Genetics and Genomics for Personalized Medicine
Edited by Il-Jin Kim

ISBN 978-981-4669-87-0 (Hardcover), 978-1-315-36476-6 (eBook)
www.panstanford.com

and 695,000 worldwide in 2008, respectively.[1] It is ranking the third cause of cancer-related death globally.[2] During recent 20 years, liver cancer showed drastic increase in death rate from 463,000 to 752,000 (62.4%) in 2010. Although the most cases occur in East Asia especially China and in Middle and Western Africa, liver cancer has been regarded as an alarming cancer even in United States and Europe, previously thought to be a low endemic regions. It is the most increasing cancer and cause of cancer-related death in those regions,[1,3,4] and is expected to be doubled over the next 10 to 20 years.[3–5]

3.1.2 Clinical Characteristics of Liver Cancer

Primary liver cancer is a tumor arisen from the liver. Hepatocellular carcinoma (HCC), which originates from hepatocytes, is the most common form, accounting for 75% of liver cancer. The second is intrahepatic cholangiocarcinoma, originating from the bile duct. The other cancers form from blood vessels, immune cells, and other mesenchymal cells in the liver. Primary liver cancer, especially HCC, is characterized by its well-known carcinogenic causes. Chronic infection of hepatitis B virus (HBV) is a major cause of liver cancer in Asia and sub-Saharan Africa. Cirrhosis or chronic hepatitis from hepatitis C virus (HCV), alcohol drinking, nonalcoholic fatty liver disease, and other immune-mediated liver injury are well-known causes of liver cancer in United States, Europe, and Japan. Because most liver cancer occurs in patients at high risk, it might be a good strategy to adopt cancer surveillance for earlier detection of liver cancer. Many countries are conducting surveillance program using ultrasonography and or serum biomarkers (alpha-fetoprotein) every 6 to 12 months. However, in a majority of patients the cancer is still being detected in advanced stages despite such efforts.

Another characteristic of patients with liver cancer (HCC) is liver cirrhosis. It is a terminal stage of advanced fibrosis from chronic hepatitis and a potent precancerous state. About 80–90% of patients also have liver cirrhosis.[6] The patients suffer from decreased liver function and portal hypertension. Jaundice, encephalopathy, decreased serum albumin, deterioration of blood coagulation system, and various metabolic changes are common complications arising from decreased liver function. Portal hypertension provokes ascites, peripheral edema, and gastroesophageal varices (protruding tortuous veins). Therefore, the coexistence of liver cancer and cirrhosis complicates the prognostic prediction and therapeutic strategies.[7] Even in the early stage, curative treatment could be impossible because of underlying liver cirrhosis. Liver transplantation is only an option for the patient. For advanced liver cancer, unpredicted toxicity has been a major hurdle in many clinical trials of new multikinase inhibitors in patients with liver cirrhosis.[8]

High recurrence rate is another challenge in patients under treatment. A 5-year recurrence rate was more than 70% in patients with curative hepatic resection. Early recurrence occurs usually within 2 years of treatment, usually from remnant tumor (micrometastasis). Late recurrence occurs in other sites of the liver after 2 years and it is regarded as a de novo carcinogenesis in the precancerous liver. More detailed prediction models and various chemoprevention strategies are under development in patients undergoing curative treatments. For patients with advanced liver cancer, it is challenging that there is no proven systemic therapy except for sorafenib. Even though it showed beneficial effect on patients with advanced HCC,[9] the effect is marginal and only a minority of patients showed dramatic response.

3.2 Why Personalized Medicine is Important in Patients with Liver Cancer?

Most liver cancer occurs in a chronically damaged liver with fibrosis. Several deregulated signaling pathways and genetic mutations accumulate in the damaged liver.[8] This heterogeneity of liver cancer is closely correlated with unpredictable or poor outcomes of systemic therapy.

Sorafenib, an oral multikinase inhibitor, is an only proven systemic agent to increase overall survival in patients with advanced HCC.[9] However, its antineoplastic effect is marginal, suggesting that only a small fraction of patients might respond to the treatment. Recently, extensive search for informative biomarkers has been taking place to identify patients who would get greater benefits from sorafenib treatment.[10–13] However, no robust markers have emerged from screening yet. Thus, selecting an optimal candidate for sorafenib treatment is a challenge to many clinicians. Until 2013, most phase III clinical trials with multikinase inhibitors, including sunitinib, brivanib, linifanib, and erlotinib, have failed in patients with advanced HCC.[14–17] The major reason of the failure is not clear; however, genetic heterogeneity of liver cancer and lack of understanding of critical drivers of tumor progression might be accountable for the failure.[8] In a phase II trial, tivantinib (a selective oral inhibitor of MET) showed no effect on liver cancer, but a significant survival benefit was achieved when MET-positive patients were included.[18] Large-scale mutational screening approaches have enabled the identification of new disease drivers in some solid tumors such as lung, breast, or melanoma.[19] This approach will open the opportunity to identify underlying drivers in liver cancer, too. Molecular classification of liver cancer would have a greater role in future rationalized clinical trials and will enable identification of the patients who will benefit from certain treatments. Recent development in technology have shown that personalized medicine is not far from becoming a reality.

3.3 Methods and Results of Genomic Profiling of Liver Cancer

Vast information on liver cancer has accumulated from genome-wide approaches, including copy number analysis, gene expression profiling, methylation profiling, and whole-genome sequencing.

3.3.1 Comparative Genomic Hybridization (CGH)

This technique was first developed to compare chromosomal differences more efficiently between solid tumor and normal tissue.[20] Because cancer genomes have diverse genetic rearrangements and copy number aberration, CGH has become a popular tool for cancer research. The resolution of traditional CGH was very limited at first, but it has been improved to two megabases. Combining the DNA microarray method, CGH techniques could increase the resolution as low as a few kilobases.[21] In liver cancer, multiple studies have reported prevalent amplification of 1q, 6p, 8q, and 17q regions and frequent deletion of 4q, 8p, 16q, and 17p regions.[22–26] The aberration patterns did not differ between HCCs from chronic hepatitis B or C.[26] The patterns were implicated with antiviral immunity pathways.[26] Among the amplified regions, 8q24 loci have been frequently reported in many studies, in which MYC is located.[22–25]

3.3.2 Microarray-Based Technology

Using microarray technology, we can evaluate genetic or proteomic alterations of specific cells or tissues extensively in a simultaneous manner. Microscopic arrays containing a number of probes are attached on a solid plate or beads and probe-target hybridization is measured in parallel. This technology was first introduced to assess expression of mRNAs or non-coding RNAs. Gene expression profiling studies for many malignant tumors have been reported using microarray technologies.[27–29] These approaches have identified many conserved gene expression signatures related to new therapeutic targets or to predicting prognosis in terms of overall survival or recurrence-free survival rates more accurately than conventional staging systems in various cancers.

Moreover, microarray technology is being widely used as a key method in other studies, such as uncovering single-nucleotide polymorphisms associated with cancer risk, methylation status of gene promoters, DNA copy number alterations, protein expression profiling, and re-sequencing of cancer genomes. The microarray technology combined with big data processing might be an optimal method to investigate the complex systems of deregulated cell biology.

In previous studies,[30–34] an unbiased analytical approach applied to gene expression data from human HCC identified distinct subtypes of HCC significantly associated with patient survival. These findings suggest that gene expression profiling signatures accurately reflect biological and clinical differences between subtypes of HCC and would be highly valuable in determining patient prognosis. The current clinical challenge is to identify patients who do not derive much benefit from conventional therapies and to offer alternative treatments. If key (or master) regulators (genes, pathways, and/or networks) driving the biology of the tumor can be identified, they might lend themselves to therapeutic exploitation. In this context, it is not enough to rely entirely on gene expression signatures that are indicative of prognosis, since the profiles may fall short of explaining at the molecular level what drives the prognostic difference between subtypes of tumors.

Protein microarrays have been developed by adopting the knowledge and technical innovations that have made DNA microarrays possible. The technical aspects of miniaturizing traditional methods, such as western blotting and protein dotting onto nitrocellulose or nylon membranes, were quickly adapted to protein microarray technology. In reverse-phase protein array (RPPA) approach, tissue lysates are immobilized in solid surface and an array comprises thousands of different patient samples. Each array is then incubated with one antibody, and a protein feature is measured and directly compared across multiple samples.[35–38] Although there are no major proteomic studies of liver cancer using RPPA technology yet, it is anticipated that it would provide highly informative data that can uncover potential druggable targets because a vast majority of antibody probes are pre-selected for well-characterized signaling proteins including kinases with available potent inhibitors.

3.3.3 Next-Generation Sequencing

Growing demand for more efficient DNA sequencing gave rise to new technologies, named next-generation sequencing (NGS). They were developed on the basis of new biotechnology and computing systems that can control complex and big data. NGS includes various sequencing methods to read a large volume of sequence in faster and inexpensive manners. Many different companies have developed different methods, such as Roche Applied Science (454 Genome Sequencer FLX System, Indianapolis, IN, USA), Life Technologies (Sequencing by Oligonucleotide Ligation and Detection or SOLiD, Carlsbad, CA, USA), Illumina (Genome Analyzer II, San Diego, CA, USA), Helicos BioSciences (HeliScope Single Molecule Sequencer,

Cambridge, MA, USA), and Ion Torrent Systems (now owned by Life Technologies, Ion Proton System, Carlsbad, CA, USA).

More accurate and cost-effective sequencing methods are still under development. These new technologies provided unique opportunities to investigate all sequences of entire cancer genomes and to understand how genetic differences affect disease.[39]

Recently, comprehensive genome sequencing for whole genome or exome has been conducted in liver cancers by many researchers.[40–46] The results confirmed again that liver cancer is very heterogeneous in genetic alterations.[47] The median average of mutation rate was about 2∼4 per megabase. However, the rage of mutation rate was very diverse among the tumors (∼1 to more than 300 per tumor).[44–46] Genetic signatures were different among the patients, multicentric tumors in a patient, and even in a same tumor.[41,42,47] Despite severe genetic heterogeneity, a few common signaling pathways were refined, including Wnt/β-catenin pathway and cell cycle regulatory pathway (TP53 and CDKN2A).[40,42–46] Next-generation studies have identified new carcinogenic pathways, epigenetic modifier (ARID and MLL family),[40,42,44,45] stress oxidative pathway (NRF2, KEAP1),[43,45] and JAK/STAT pathway.[46] Although it provides great possibility to reveal carcinogenesis, NGS is still in an early stage in human liver cancer compared with other cancers (breast, lung, colon, and melanoma). NGS is challenged not only by great heterogeneity of the tumor, but also by vast genomic data including considerable unknown information. We don't know whether they are valuable or simply nonsensical mutations. As more data are accumulating and innovative analytic methods are developing, NGS might be a key tool for elucidating key oncogenic pathways among the many heterogeneous genetic aberrations in human liver cancer.

3.3.4 Integromics: Integration of Multiple -omic Data

The purpose of integromics in liver cancer is to reveal different carcinogenic pathways and to guide us to overview the development of liver cancer. Most liver cancers have vast structural or functional changes in genome, transcripts, and proteins. A majority of them might be simply reactive changes in carcinogenesis. It is not easy to find carcinogenic addiction loops (driver gene) from the complicated mess of echo changes (passenger genes). Therefore, the effort to combine multiple data sets, including somatic mutations, DNA copy number alterations, promoter methylation, and expression of coding and non-coding RNAs, has recently emerged. The major drivers of liver cancer will be identified and classified in the near future by integration of these multiple -omic data.

The Cancer Genome Atlas (TCGA) Project is a landmark research program supported by the National Human Genome Research Institute and National Cancer Institute at the National Institutes of Health. The goal of TCGA is to collect comprehensive genomic and proteomic information on all major human cancers, including liver cancer. Initial efforts focused on glioblastoma, lung squamous cell cancer, and ovarian cancer as pilot projects.[48–51] By using various different platforms, TCGA currently gathers many different genome-wide data, including mRNA expression, microRNA expression, somatic mutations, copy number alteration, and promoter methylation. In addition, it also generates proteomic data by using RPPA technology. The project plans to collect genomic and proteomic data from more than 500 tissues per cancer type and release the data to public without any restriction in use of the data.

This ambitious project has identified novel driver genes and biomarkers on the basis of genomic, transcriptomic, proteomic, and epigenomic alterations. Some of findings are clinically relevant and unexpected. For example, we now learned that non-hypermutated adenocarcinomas of the colon and rectum are not distinguishable at the genomic level.[52] In lung squamous cell cancer, while KRAS and EGFR mutations, most commonly activated oncogenes in lung adenocarcinoma, are extremely rare, alterations in the FGFR kinase family are common.[51] Thus, massive data from a large number of tissues created unprecedented opportunity for taking an integrated approach toward a systems-level understanding of disruptions in cellular and molecular pathways in cancer. The results of the project for liver cancer will be open to public soon.

The global effort to combine and analyze the vast data from genomic and proteomic research will make a list of common carcinogenic driver pathways for liver cancer in the near future. Personalized medicine will be possible for most patients with liver cancer in selecting appropriate treatment strategy based on their individual genomic and proteomic data sets.

3.4 Conclusion

Personalized medicine is a promising opportunity to many patients and physicians. However, there are no clinically recognized molecular subtypes in liver cancer yet, indicating that there is still a wide gap between concept and real practice. To overcome the current obstacles, more detailed profiling of driver pathways is needed. To explore the hepatocarcinogenesis, global and multidisciplinary cooperation will be intensified using innovated and powerful bioinformatics. In the near future, each patient with liver cancer will take a test for individual carcinogenic pathway. On the basis of the result, the optimal treatment strategy for one or more specific driver pathways will be given to the patient.

References

1. Jemal A, Bray F, Center MM, Ferlay J, Ward E, Forman D. (2011). Global cancer statistics. *CA Cancer J Clin*, **61**, 69–90.
2. Lozano R, Naghavi M, Foreman K, Lim S, Shibuya K, Aboyans V, et al. (2012). Global and regional mortality from 235 causes of death for 20 age groups in 1990 and 2010: a systematic analysis for the Global Burden of Disease Study 2010. *Lancet*, **380**, 2095–2128.
3. El-Serag HB. (2004). Hepatocellular carcinoma: recent trends in the United States. *Gastroenterology*, **127**, S27–S34.
4. El-Serag HB, Mason AC. (1999). Rising incidence of hepatocellular carcinoma in the United States. *N Engl J Med*, **340**, 745–750.
5. Altekruse SF, McGlynn KA, Reichman ME. (2009). Hepatocellular carcinoma incidence, mortality, and survival trends in the United States from 1975 to 2005. *J Clin Oncol*, **27**, 1485–1491.
6. Fattovich G, Stroffolini T, Zagni I, Donato F. (2004). Hepatocellular carcinoma in cirrhosis: incidence and risk factors. *Gastroenterology*, **127**, S35–50.

7. Villanueva A, Hernandez-Gea V, Llovet JM. (2013). Medical therapies for hepatocellular carcinoma: a critical view of the evidence. *Nat Rev Gastroenterol Hepatol*, **10**, 34–42.
8. Llovet JM, Hernandez-Gea V. (2014). Hepatocellular carcinoma: reasons for phase III failure and novel perspectives on trial design. *Clin Cancer Res*, **20**, 2072–2079.
9. Llovet JM, Ricci S, Mazzaferro V, Hilgard P, Gane E, Blanc JF, et al. (2008). Sorafenib in advanced hepatocellular carcinoma. *N Engl J Med*, **359**, 378–390.
10. Lee JH, Park JY, Kim do Y, Ahn SH, Han KH, Seo HJ, et al. (2011). Prognostic value of 18F-FDG PET for hepatocellular carcinoma patients treated with sorafenib. *Liver Int*, **31**, 1144–1149.
11. Llovet JM, Pena CE, Lathia CD, Shan M, Meinhardt G, Bruix J, et al. (2012). Plasma biomarkers as predictors of outcome in patients with advanced hepatocellular carcinoma. *Clin Cancer Res*, **18**, 2290–2300.
12. Bargellini I, Scionti A, Mismas V, Masi G, Vivaldi C, Bartolozzi C, et al. (2014). Identification of responders to sorafenib in hepatocellular carcinoma: is tumor volume measurement the way forward? *Oncology*, **86**, 191–198.
13. Horwitz E, Stein I, Andreozzi M, Nemeth J, Shoham A, Pappo O, et al. (2014). Human and mouse VEGFA-amplified hepatocellular carcinomas are highly sensitive to sorafenib treatment. *Cancer Discov*, **4**, 730–743.
14. Cainap C, Qin SK, Huang WT, Chung IJ, Pan HM, Cheng Y, et al. (2013). Phase III trial of linifanib versus sorafenib in patients with advanced hepatocellular carcinoma (HCC). *J Clin Oncol*, **31**, 1.
15. Cheng A-L, Kang Y-K, Lin D-Y, Park J-W, Kudo M, Qin S, et al. (2013). Sunitinib versus sorafenib in advanced hepatocellular cancer: results of a randomized phase III trial. *J Clin Oncol*, **31**, 4067–4075.
16. Johnson PJ, Qin S, Park JW, Poon RT, Raoul JL, Philip PA, et al. (2013). Brivanib versus sorafenib as first-line therapy in patients with unresectable, advanced hepatocellular carcinoma: results from the randomized phase III BRISK-FL study. *J Clin Oncol*, **31**, 3517–3524.
17. Zhu AX, Rosmorduc O, Evans TR, Ross PJ, Santoro A, Carrilho FJ, et al. (2015). SEARCH: a phase III, randomized, double-blind, placebo-controlled trial of sorafenib plus erlotinib in patients with hepatocellular carcinoma. *J Clin Oncol*, **33**, 559–566.
18. Santoro A, Rimassa L, Borbath I, Daniele B, Salvagni S, Van Laethem JL, et al. (2013). Tivantinib for second-line treatment of advanced

hepatocellular carcinoma: a randomised, placebo-controlled phase 2 study. *Lancet Oncol*, **14**, 55–63.

19. Villanueva A, Llovet JM. (2014). Liver cancer in 2013: mutational landscape of HCC–the end of the beginning. *Nat Rev Clin Oncol*, **11**, 73–74.
20. Kallioniemi A, Kallioniemi OP, Sudar D, Rutovitz D, Gray JW, Waldman F, et al. (1992). Comparative genomic hybridization for molecular cytogenetic analysis of solid tumors. *Science*, **258**, 818–821.
21. Pinkel D, Albertson DG. (2005). Array comparative genomic hybridization and its applications in cancer. *Nat Genet*, **37 Suppl**, S11–17.
22. Zimonjic DB, Keck CL, Thorgeirsson SS, Popescu NC. (1999). Novel recurrent genetic imbalances in human hepatocellular carcinoma cell lines identified by comparative genomic hybridization. *Hepatology*, **29**, 1208–1214.
23. Poon TC, Wong N, Lai PB, Rattray M, Johnson PJ, Sung JJ. (2006). A tumor progression model for hepatocellular carcinoma: bioinformatic analysis of genomic data. *Gastroenterology*, **131**, 1262–1270.
24. Moinzadeh P, Breuhahn K, Stutzer H, Schirmacher P. (2005). Chromosome alterations in human hepatocellular carcinomas correlate with aetiology and histological grade–results of an explorative CGH meta-analysis. *Br J Cancer*, **92**, 935–941.
25. Kusano N, Shiraishi K, Kubo K, Oga A, Okita K, Sasaki K. (1999). Genetic aberrations detected by comparative genomic hybridization in hepatocellular carcinomas: their relationship to clinicopathological features. *Hepatology*, **29**, 1858–1862.
26. Guo X, Yanna, Ma X, An J, Shang Y, Huang Q, et al. (2011). A meta-analysis of array-CGH studies implicates antiviral immunity pathways in the development of hepatocellular carcinoma. *PLoS One*, **6**, e28404.
27. Bittner M, Meltzer P, Chen Y, Jiang Y, Seftor E, Hendrix M, et al. (2000). Molecular classification of cutaneous malignant melanoma by gene expression profiling. *Nature*, **406**, 536–540.
28. Beer DG, Kardia SL, Huang CC, Giordano TJ, Levin AM, Misek DE, et al. (2002). Gene-expression profiles predict survival of patients with lung adenocarcinoma. *Nat Med*, **8**, 816–824.
29. van de Vijver MJ, He YD, van't Veer LJ, Dai H, Hart AA, Voskuil DW, et al. (2002). A gene-expression signature as a predictor of survival in breast cancer. *N Engl J Med*, **347**, 1999–2009.
30. Ye QH, Qin LX, Forgues M, He P, Kim JW, Peng AC, et al. (2003). Predicting hepatitis B virus-positive metastatic hepatocellular carcinomas using

gene expression profiling and supervised machine learning. *Nat Med*, **9**, 416–423.

31. Lee JS, Chu IS, Heo J, Calvisi DF, Sun Z, Roskams T, et al. (2004). Classification and prediction of survival in hepatocellular carcinoma by gene expression profiling. *Hepatology*, **40**, 667–676.
32. Lee JS, Heo J, Libbrecht L, Chu IS, Kaposi-Novak P, Calvisi DF, et al. (2006). A novel prognostic subtype of human hepatocellular carcinoma derived from hepatic progenitor cells. *Nat Med*, **12**, 410–416.
33. Hoshida Y, Villanueva A, Kobayashi M, Peix J, Chiang DY, Camargo A, et al. (2008). Gene expression in fixed tissues and outcome in hepatocellular carcinoma. *N Engl J Med*, **359**, 1995–2004.
34. Woo HG, Park ES, Cheon JH, Kim JH, Lee JS, Park BJ, et al. (2008). Gene expression-based recurrence prediction of hepatitis B virus-related human hepatocellular carcinoma. *Clin Cancer Res*, **14**, 2056–2064.
35. Espina V, Mehta AI, Winters ME, Calvert V, Wulfkuhle J, Petricoin EF, 3rd, et al. (2003). Protein microarrays: molecular profiling technologies for clinical specimens. *Proteomics*, **3**, 2091–2100.
36. Liotta LA, Espina V, Mehta AI, Calvert V, Rosenblatt K, Geho D, et al. (2003). Protein microarrays: meeting analytical challenges for clinical applications. *Cancer Cell*, **3**, 317–325.
37. Haab BB. (2005). Antibody arrays in cancer research. *Mol Cell Proteomics*, **4**, 377–383.
38. Sheehan KM, Calvert VS, Kay EW, Lu Y, Fishman D, Espina V, et al. (2005). Use of reverse phase protein microarrays and reference standard development for molecular network analysis of metastatic ovarian carcinoma. *Mol Cell Proteomics*, **4**, 346–355.
39. Metzker ML. (2010). Sequencing technologies - the next generation. *Nat Rev Genet*, **11**, 31–46.
40. Li M, Zhao H, Zhang X, Wood LD, Anders RA, Choti MA, et al. (2011). Inactivating mutations of the chromatin remodeling gene ARID2 in hepatocellular carcinoma. *Nat Genet*, **43**, 828–829.
41. Totoki Y, Tatsuno K, Yamamoto S, Arai Y, Hosoda F, Ishikawa S, et al. (2011). High-resolution characterization of a hepatocellular carcinoma genome. *Nat Genet*, **43**, 464–469.
42. Fujimoto A, Totoki Y, Abe T, Boroevich KA, Hosoda F, Nguyen HH, et al. (2012). Whole-genome sequencing of liver cancers identifies etiological influences on mutation patterns and recurrent mutations in chromatin regulators. *Nat Genet*, **44**, 760–764.

43. Guichard C, Amaddeo G, Imbeaud S, Ladeiro Y, Pelletier L, Maad IB, et al. (2012). Integrated analysis of somatic mutations and focal copy-number changes identifies key genes and pathways in hepatocellular carcinoma. *Nat Genet*, **44**, 694–698.

44. Huang J, Deng Q, Wang Q, Li KY, Dai JH, Li N, et al. (2012). Exome sequencing of hepatitis B virus-associated hepatocellular carcinoma. *Nat Genet*, **44**, 1117–1121.

45. Cleary SP, Jeck WR, Zhao X, Chen K, Selitsky SR, Savich GL, et al. (2013). Identification of driver genes in hepatocellular carcinoma by exome sequencing. *Hepatology*, **58**, 1693–1702.

46. Kan Z, Zheng H, Liu X, Li S, Barber TD, Gong Z, et al. (2013). Whole-genome sequencing identifies recurrent mutations in hepatocellular carcinoma. *Genome Res*, **23**, 1422–1433.

47. Nakagawa H, Shibata T. (2013). Comprehensive genome sequencing of the liver cancer genome. *Cancer Lett*, **340**, 234–240.

48. Cancer Genome Atlas Research N. (2008). Comprehensive genomic characterization defines human glioblastoma genes and core pathways. *Nature*, **455**, 1061–1068.

49. International Cancer Genome C, Hudson TJ, Anderson W, Artez A, Barker AD, Bell C, et al. (2010). International network of cancer genome projects. *Nature*, **464**, 993–998.

50. Cancer Genome Atlas Research N. (2011). Integrated genomic analyses of ovarian carcinoma. *Nature*, **474**, 609–615.

51. Cancer Genome Atlas Research N. (2012). Comprehensive genomic characterization of squamous cell lung cancers. *Nature*, **489**, 519–525.

52. Cancer Genome Atlas N. (2012). Comprehensive molecular characterization of human colon and rectal cancer. *Nature*, **487**, 330–337.

Chapter 4

Applications of Circulating DNA Analysis in Personalized Medicine

Dana W. Y. Tsui[a] **and Muhammed Murtaza**[b,c]

[a] *Center for Molecular Oncology, Department of Pathology, Memorial Sloan Kettering Cancer Center, New York, USA*
[b] *Center for Noninvasive Diagnostics, Translational Genomics Research Institute, 445 N 5th St, Phoenix, AZ, USA*
[c] *Mayo Clinic, 13400 E Shea Blvd, Scottsdale, AZ 85259, USA*
mmurtaza@tgen.org, tsuiw@mskcc.org

Extracellular fragments of DNA circulating in plasma and serum have been recognized for over six decades. Recent advances in molecular techniques and massively parallel sequencing have enabled exciting applications of circulating DNA analysis for noninvasive personalized diagnostics. In this chapter, we provide an overview of contemporary analytical methods and recent advances in applications.

4.1 Biological Characteristics of Circulating DNA

4.1.1 History

The first description of circulating DNA in humans dates back to 1948 by Mendel and Métais (Mendel et al., 1948). Since then,

Cancer Genetics and Genomics for Personalized Medicine
Edited by Il-Jin Kim

ISBN 978-981-4669-87-0 (Hardcover), 978-1-315-36476-6 (eBook)
www.panstanford.com

aberrant levels of circulating DNA have been found in the plasma and serum of patients with different physiological conditions (Tan et al., 1966; Koffler et al., 1973; Leon et al., 1977; Stroun et al., 1977), including tumor-derived DNA in cancer patients (Chen et al., 1996; Nawroz et al., 1996; Sorenson et al., 1994) and fetal-derived DNA in pregnant women (Lo et al., 1997). These findings highlighted tremendous opportunities for noninvasive diagnostics in diseases such as cancer and fetal disorders.

4.1.2 Biological Characteristics

Little is known about the origin and biological characteristics of circulating DNA. Growing evidence suggests that tumor-derived DNA is released into the circulation during apoptotic and necrotic cell death (Jahr et al., 2001; Rainer et al., 2001; Holdenrieder et al., 2005). This is supported by the observation that cell-free plasma DNA is generally fragmented to less than 200 bp in length (Chan et al., 2004; Mouliere et al., 2013), with a characteristic 10 bp step in the size distribution suggesting enzymatic degradation such as nuclease-cleaved nucleosomes (Fan et al., 2010; Lo et al., 2010). Circulating DNA fragments carrying fetal or cancer-specific genomic alterations tend to be shorter than host circulating DNA (Yu et al., 2014; Jiang et al., 2015).

In terms of kinetics, circulating DNA exhibits rapid turnover in the blood stream. Fetal-derived DNA in maternal plasma is cleared within 1–2 days after the delivery of the fetus, with an initial half-life of 1 hour (Yu et al., 2013). The half-life of tumor-derived DNA following surgical resection of colorectal cancer was observed to be 2 hours (Diehl et al., 2008). Circulating DNA is, at least in part, cleared through the kidney as evident by detection of fetal and tumor-derived DNA in urine (Botezatu et al., 2000). Postpartum serial analysis revealed no evidence of progressive decrease in size of fetal DNA in maternal plasma after delivery, suggesting fetal DNA clearance may involve additional mechanisms with no preference for fragment size (Yu et al., 2013).

Apart from cancer and fetal diseases, aberrations in total circulating DNA levels have also been documented in other physiological conditions such as systemic lupus erythematosus

(Raptis and Menard 1980; Chan et al., 2014), trauma (Lo et al., 2000; Yamanouchi et al., 2013), and organ failure in critically ill patients (Rhodes et al., 2006; Saukkonen et al., 2008). Improved understanding of biological characteristics of circulating DNA can guide design of molecular assays for diverse applications of noninvasive diagnostics.

4.2 Molecular Methods for Circulating DNA Analysis

The choice of an ideal method for circulating DNA analysis is driven by interplay between technical accuracy, quantitative performance, and allelic discrimination of an assay at any given genomic locus, the type of genomic alteration targeted, and the number of genomic loci assayed.

Detection of circulating DNA can be achieved using PCR-based methods such as restriction fragment length polymorphism analysis, allele-specific PCR, or chain-termination ("Sanger") sequencing. These vary in their sensitivity for low-abundance mutations, do not provide quantitative results and performance is dependent on assay design.

In earlier studies, locus-specific quantitative PCR (qPCR) assays were used to infer total cell-free DNA levels. These assays rely on PCR primers flanking a genomic locus with either a double-stranded DNA dye or a fluorescent-labeled oligonucleotide probe. The accuracy of qPCR is assay dependent and its performance is adequate for total circulating DNA quantification. Quantitative PCR may also enable investigation of relative copy number changes and somatic structural variants with known breakpoints, although this requires appropriate molecular standards for comparison. Identification of point mutations in circulating DNA is possible, but the sensitivity of qPCR assays for low-abundance alleles expected in circulating DNA is limited. The number of genomic loci that can be assayed in multiplex remains limited.

Identification and quantification of mutant alleles can also be performed using matrix-assisted laser desorption ionization time-of-flight (MALDI-TOF) mass spectrometry. MALDI-TOF allows high-throughput analysis of up to ~100 single genomic loci with reliable results when allele fractions are greater than 10% in input DNA (Gut, 2004).

Digital PCR relies on segregation of template DNA into many hundreds or thousands of PCR reactions with one or no copies of the targeted fragment (Vogelstein and Kinzler, 1999). The presence of DNA template, usually detected by fluorescence, is measured for each reaction, either in real-time or at endpoint, to count the number of positive reactions. Unlike qPCR that requires comparison with a standard input, digital PCR allows direct counting of the absolute number of input molecules. As each positive PCR reaction amplifies one template molecule, digital PCR using mutation-specific fluorescent probes is particularly useful for quantification of low-abundance alleles in a sample. It allows absolute sensitivity for detection of mutant DNA fragments, limited only by number of digital replicates and background noise for each PCR assay. In practice, digital PCR is highly sensitive for point mutations with allele fractions as low as 0.01%. Although digital PCR can be performed by diluting input DNA material and setting up digital reactions in 96- or 384-well plates, several high-throughput methods have been described, including microfluidic devices (Yung et al., 2009), droplet-based platforms (Pinheiro et al., 2012; Taly et al., 2013), and BEAMing (beads, emulsion, amplification, and magnetics) (Diehl et al., 2008). The number of genomic loci that can be assayed in multiplex remains limited to a few.

Advances in massively parallel sequencing (MPS) have enabled great strides in circulating DNA research. Whole genome sequencing (WGS) of circulating DNA is available as a clinical test for screening of fetal aneuploidies from maternal plasma. Sequencing libraries for WGS are prepared using ligation of universal sequencing adaptors to flank cell-free DNA fragments. WGS allows hypothesis-free analysis of circulating DNA but at current sequencing costs, the depth of coverage practically achievable (~30-fold) limits technical accuracy at any one genomic locus. As a result, WGS is more suited for aggregate analysis across several genomic loci (copy number alterations [Chan et al., 2013], structural variation [Leary et al.,

2012], or quantification of donor-derived DNA [Snyder et al., 2011]) than individual point mutations. Whole exome sequencing (WES) relies on enrichment of sequencing libraries through hybridization with oligonucleotides targeting annotated exonic regions. WES allows greater depth of coverage (typically ~100–200-fold) and identification and reliable quantification of point mutations with allele fractions >5% of circulating DNA. WES remains a useful tool for relatively unbiased identification of somatic mutations in ctDNA obtained from advanced cancer patients who present with high disease burden (Murtaza et al., 2013).

Greater depth of coverage (~1000s-fold) can be achieved by focusing on a few genes of interest using custom enrichment by hybridization (Newman et al., 2014) or multiplexed PCR-based amplicon sequencing (Forshew et al., 2012). Both methods are accurate for identification of mutations with >1–2% allele fraction and quantification of known mutations with >0.1% allele fraction. Enrichment by hybridization allows several megabases of the genomic coverage compared with several kilobases of coverage achievable with current approaches for amplicon sequencing. However, ligation-based library preparation methods are required for enrichment by hybridization and these have limited efficiency with <10 nanograms of input DNA and can suffer from allelic dropout. In contrast, amplicon-based methods such as tagged amplicon deep dequencing (TAm-Seq) have been shown to prepare sequencing libraries from as little as 1–2 genome equivalents of input DNA (Forshew et al., 2012).

Improved sensitivity for direct identification of mutations with allele fractions <2% has been demonstrated using error-suppressed sequencing or using mutation enrichment prior to sequencing. Error-suppressed sequencing relies on improvements in variant calling (Gerstung et al., 2012) and molecular template barcoding (Kinde et al., 2011). Mutation enrichment is performed by modifying PCR conditions (such as denaturation temperature [Li et al., 2008] or oligonucleotide suppression of reference allele [Hyman et al., 2015]) to selectively amplify mutant DNA fragments or removal of wild-type DNA fragments by physical methods (Thompson et al., 2012). These approaches target one or a few genomic loci in multiplex or have limited efficiency for low input amounts of DNA.

4.3 Circulating Tumor-Specific DNA in Cancer Patients

Total levels of cell-free DNA in plasma and serum are elevated in cancer patients (Leon et al., 1977). However, the diagnostic performance of total cell-free DNA is limited because of non-specificity. In cancer patients, a variable fraction of cell-free DNA is contributed by the tumor itself and carries cancer-specific somatic mutations. Analysis of circulating tumor-specific DNA (ctDNA) relies on identification, detection, and quantification of these somatic mutations in plasma.

4.3.1 Monitoring of Tumor Burden and Disease Response

Longitudinal changes in ctDNA levels in cancer patients track tumor burden and can inform treatment outcomes. Unlike other circulating biomarkers (peptides or RNA expression), ctDNA analysis relies on somatic mutations that are cancer specific. Diehl et al. identified somatic point mutations in tumor samples from 18 patients with advanced colorectal cancer and tracked these mutations in serial plasma samples using BEAMing, a digital PCR-like approach that allows accurate quantification of known somatic mutations (Diehl et al., 2008). ctDNA levels tracked with clinical and radiological disease progression in patients receiving multi-modality treatment. Clearance of ctDNA post-treatment was predictive of survival at approximately 2 years of clinical follow-up. Dawson et al. investigated ctDNA in serial plasma samples from 30 patients with metastatic breast cancer receiving systemic treatment and detected patient-specific somatic point mutations in ctDNA in 29 patients (Dawson et al., 2013). Serial changes in ctDNA showed higher dynamic range and greater correlation with changes in tumor burden than CA 15-3 or the number of circulating tumor cells detectable using CellSearch. Pre-treatment ctDNA levels were informative of overall survival. Similar results have been observed in additional studies of advanced lung cancer (Newman et al., 2014; Oxnard et al., 2014), osteosarcoma (McBride et al., 2010), and melanoma (Lipson et al., 2014).

More recently, ctDNA analysis has been explored in patients with localized curable cancers. Bettegowda et al., performed an extensive survey of 640 patients with localized and metastatic non-hematological cancers (Bettegowda et al., 2014). They quantified ctDNA levels using digital genomic methods targeting one patient-specific somatic alteration in a plasma sample from each patient. ctDNA was detectable in 55% of all localized cases. Beaver et al. reported pre-operative detection of PIK3CA mutations in ctDNA using droplet digital PCR in 14/15 localized breast cancer patients where the corresponding mutation was present in the tumor (Beaver et al., 2014). In 5 of 10 patients, ctDNA remained detectable following tumor resection and while clinical follow-up was limited, 1 of these 5 patients developed clinical recurrence 26 months post diagnosis.

4.3.2 Molecular Stratification for Targeted Therapies

Recent advances in molecularly targeted cancer therapy have improved outcomes in multiple cancer types. Patients are selected for treatment based on the presence of actionable somatic mutations in tumor biopsies. Intra-tumor molecular heterogeneity limits how much a single tumor biopsy reflects the systemic tumor (Gerlinger et al., 2012). Multiple biopsies have been advocated but remain impractical in clinical practice. ctDNA analysis in patients with metastatic cancers can allow identification of actionable somatic mutations, enabling patient selection for targeted treatments. This noninvasive "liquid biopsy" approach can complement and potentially circumvent the need for invasive tissue re-biopsies.

Several retrospective studies have evaluated noninvasive molecular stratification by ctDNA analysis. Bettegowda et al. sequenced KRAS codon 12 in ctDNA from 206 patients with metastatic colorectal cancer and found 88.2% sensitivity with 99.2% specificity for noninvasive mutation detection (Bettegowda et al., 2014). Thierry et al. found >96% concordance for KRAS and BRAF hotspot mutations in 105 patients with metastatic colorectal cancer (Thierry et al., 2014). Lebofsky et al. sequenced ctDNA for actionable regions in 46 genes in 34 patients with non-hematological metastatic cancer enrolled in genome-guided targeted therapeutics trial. In 27 patients with mutations identified in tumor, they reported 97% concordance (Lebofsky et al., 2014).

4.3.3 Analysis of Clonal Evolution and Therapeutic Resistance

In patients who respond to targeted therapies, the benefits are generally short-lived. Therapeutic resistance due to tumor heterogeneity and clonal evolution arises in most metastatic cancer patients, usually within 6–12 months. Tumor re-biopsies in these patients can guide the choice of subsequent treatments and inform our understanding of mechanisms that drive acquired resistance. However, in some patients, tumor biopsies from multiple sites of disease progression have revealed independent mechanisms of treatment resistance within the same patient (Shi et al., 2014). ctDNA analysis in plasma samples obtained at resistance can allow noninvasive identification of resistance mechanisms and may overcome the limitations of singletumor biopsies.

Murtaza et al. described an exome sequencing approach for comparison of pre- and post-treatment ctDNA in metastatic cancer patients receiving systemic therapy (Murtaza et al., 2013). In this proof-of-principle study of 6 patients, potential drivers of therapeutic resistance were identified and clonal evolution driven by treatment was tracked with tumor independent analysis. Bettegowda et al. performed targeted sequencing of genes involved in the mitogen-activated protein kinase pathway and found mutations arising in post-treatment ctDNA samples from 23/24 metastatic colorectal cancer patients who progressed on EGFR blockade, confirming earlier reports (Diaz et al., 2012; Misale et al., 2012; Bettegowda et al., 2014). Carreira et al. demonstrated the utility of targeted ctDNA sequencing for tracking clonal dynamics in castration-resistant prostate cancer (Carreira et al., 2014).

4.4 Circulating DNA for Noninvasive Prenatal Diagnostics

Traditionally, prenatal diagnosis of fetal genetic disorders requires invasive sampling from the fetus, such as amniocentesis and chorionic villus sampling, which are associated with an increased risk of miscarriage. The discovery of fetal-derived DNA in maternal plasma offers opportunities for safer assessment of the fetal genetic status, leading to a paradigm shift in noninvasive prenatal diagnostics.

4.4.1 Noninvasive Diagnosis of Fetal Genetic Diseases

The presence of fetal DNA in maternal plasma was first demonstrated by the discovery of paternally inherited Y-chromosome-specific DNA in maternal plasma (Lo et al., 1997). It soon led to the development of a molecular test for noninvasive determination of fetal sex (Costa et al., 2001; Honda et al., 2002; Devaney et al., 2011) and rhesus D status (Lo et al., 1998; Finning et al., 2008). Prenatal assessment of fetal sex can inform management of sex-specific fetal disorders, such as hemophilia (Tsui et al., 2011) and congenital adrenal hyperplasia (Chiu et al., 2002). Noninvasive assessment of fetal rhesus D status allows early identification of rhesus D incompatibility, a hemolytic disease of the newborn. These two applications are examples of the earliest implementation of circulating fetal DNA test in clinical use (Evans and Kilpatrick 2010; Scheffer et al., 2011; Hill et al., 2012).

Other potential applications include the diagnosis of genetic disorders such as beta-thalassemia (Chiu et al., 2002) and sickle cell anemia (Barrett et al., 2012), or pregnancy-associated complications such as pre-eclampsia (Lo et al., 1999; Ng et al., 2003), hyperemesis gravidarum (Sugito et al., 2003), and invasive placenta (Sekizawa et al., 2002). The clinical utility of these tests is yet to be evaluated in large-scale validation studies.

4.4.2 Noninvasive Prenatal Diagnosis of Down Syndrome

Down syndrome, or trisomy 21, is the most frequent fetal chromosomal aneuploidy (compatible with life) affecting around 1 in 700 pregnancies (Parker et al., 2010). In conventional practice, pregnant women are screened for Down syndrome using a combination of biochemical and imaging tests. The diagnosis is confirmed with direct sampling of fetal genetic material using amniocentesis or chorionic villus sampling (CVS). A false-positive result from the screening test can lead to unnecessary invasive sampling of the fetus. Circulating fetal-derived DNA presents an opportunity to improve the specificity for aneuploidy screening. However, fetal DNA is present at a low proportion in maternal plasma, typically around 10% for early pregnancy, coexisting with the background of maternal DNA.

Over the past decade, different strategies have been developed to improve prenatal diagnostics of Down syndrome using circulating DNA analysis (Lo and Chiu, 2007). These approaches include quantification of fetal-specific paternally inherited SNPs (Ding et al., 2004), assessment of relative allelic ratio in chromosome 21 RNA transcripts (Go et al., 2007; Lo et al., 2007), or analysis of epigenetic markers (Chim et al., 2008; Nygren et al., 2010). WGS of maternal plasma has been described by multiple groups (Chiu et al., 2008; Fan et al., 2008; Chiu et al., 2010). WGS enables comparison of DNA fragments from chromosome 21 with other regions of the genome in maternal plasma (Chiu et al., 2009). Small increases in chromosomal dosage can be detected and this approach is universally applicable without need for prior identification of fetal-specific genetic markers. This approach can also be used for other fetal chromosomal aneuploidies such as trisomy 13 and 18, with appropriate bioinformatics analysis to adjust for chromosome-specific sequencing bias (Chen et al., 2011).

4.4.3 Noninvasive Sequencing of the Fetal Genome

Beyond fetal chromosomal abnormalities, recent studies have demonstrated that it is possible to reconstruct the entire fetal genome by WGS of maternal plasma (Lo et al., 2010; Fan et al., 2012; Kitzman et al., 2012). These reports proved that fetal genomes are proportionally represented in maternal plasma across all chromosomes, providing rationale for inference of fetal chromosome 21 imbalance from the combined circulating fetal and maternal genomes (Lo et al., 2010). Fetal genome sequencing may allow prenatal detection of single-gene disorders.

4.4.4 Clinical Implementation of Prenatal Diagnosis

Large-scale clinical validation studies of maternal plasma sequencing have demonstrated sensitivity for Down syndrome at close to 99% with a false-positive rate of <1% (Chiu et al., 2011; Palomaki et al., 2011; Bianchi et al., 2012; Lo and Chiu 2012). The clinical test has been available since 2011 (Palomaki et al., 2011), primarily being offered to women identified as high-risk for Down syndrome based on traditional screening methods (Chiu et al., 2011; Palomaki et al., 2011). A recent multi-center study in predominantly low-risk women comparing traditional screening with circulating DNA WGS showed the circulating DNA test detected all cases of aneuploidies (Bianchi et al., 2014). The study showed positive predictive value of circulating DNA for trisomy 21 was 45.5 compared to 4.2% with traditional tests. Currently, the test is available in the United States and parts of Asia and Europe through commercial providers.

4.5 Circulating DNA in Transplant Recipients

In transplant recipients, a fraction of circulating DNA fragments is contributed by the transplanted organ and carries donor-specific genetic polymorphisms (Lui et al., 2002; Hung et al., 2009). Snyder et al. developed a method called "genome transplant dynamics" to use genotyping of germline polymorphisms in transplant donors and WGS of cell-free DNA in serial post-transplant plasma samples from recipients to quantify total donor-derived DNA levels (Snyder et al., 2011). De Vlaminck et al. showed that increase in donor-derived DNA correlates with acute rejection after heart transplantation, with diagnostic performance equivalent to endocardial biopsies (De Vlaminck et al., 2014).

4.6 Pre-analytical Considerations

Cell-free fetal-derived or tumor-derived DNA often constitutes less than 10% of circulating DNA. Therefore, pre-analytical factors such processing and storage can significantly affect the outcomes of

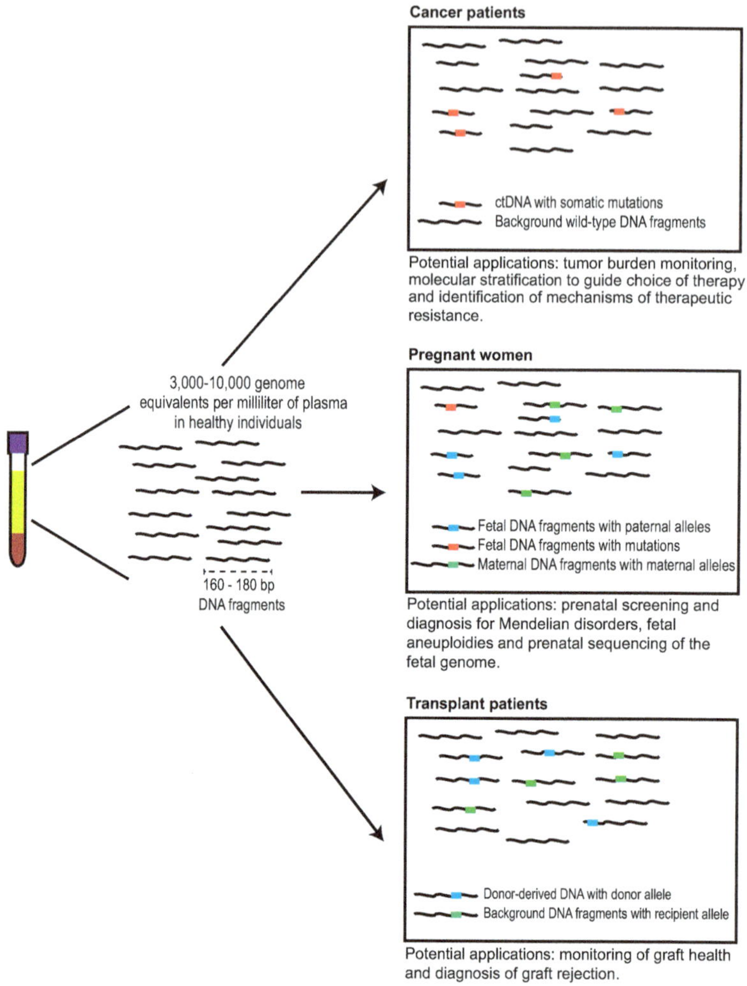

Figure 4.1

molecular analyses. Most studies described above relied on early processing of blood samples (within 1–3 hours of venipuncture) and dual centrifugation protocols to minimize the impact of background DNA from lysis of peripheral blood cells. Extensive studies have evaluated the effects of bio-fluid (such as serum vs. plasma) and processing protocols (such as blood tubes with EDTA vs. heparin, centrifugation times and speeds or delays in blood processing) on the yield and integrity of circulating DNA (Chiu et al., 2001; Lui and Dennis 2002; Barrett et al., 2011; Wong et al., 2013). Large-scale validation studies are warranted to standardize the procedures and achieve optimal sample quality for diagnostic applications.

4.7 Conclusion

Recent advances in molecular analysis methods have enabled several potential applications of circulating DNA analysis. Circulating tumor DNA can be informative as a cancer-specific personalizable biomarker and as a noninvasive liquid biopsy to guide treatment selection. Donor-derived DNA in transplant recipients can aid early diagnosis of organ rejection and improved individualized management of transplant patients. Fetal genome sequencing from maternal plasma creates new opportunities in prenatal diagnostics of genetic disorders and traits. With the exception of prenatal diagnosis for fetal aneuploidies, large-scale prospective validation studies are needed to evaluate the clinical relevance and benefit for most of these applications

References

Barrett, A. N., T. C. McDonnell, K. C. Chan, and L. S. Chitty (2012). Digital PCR analysis of maternal plasma for noninvasive detection of sickle cell anemia. *Clin Chem.* **58**(6): 1026–1032.

Barrett, A. N., B. G. Zimmermann, D. Wang, A. Holloway, and L. S. Chitty (2011). Implementing prenatal diagnosis based on cell-free fetal DNA: accurate identification of factors affecting fetal DNA yield. *PLoS One.* **6**(10): e25202.

Beaver, J. A., D. Jelovac, S. Balukrishna, R. L. Cochran, S. Croessmann, D. J. Zabransky, H. Y. Wong, P. Valda Toro, J. Cidado, B. G. Blair, D. Chu, T. Burns, M. J. Higgins, V. Stearns, L. Jacobs, M. Habibi, J. Lange, P. J. Hurley, J. Lauring, D. A. VanDenBerg, J. Kessler, S. Jeter, M. L. Samuels, D. Maar, L. Cope, A. Cimino-Mathews, P. Argani, A. C. Wolff, and B. H. Park (2014). Detection of cancer DNA in plasma of patients with early-stage breast cancer. *Clin Cancer Res.* **20**(10): 2643–2650.

Bettegowda, C., M. Sausen, R. J. Leary, I. Kinde, Y. Wang, N. Agrawal, B. R. Bartlett, H. Wang, B. Luber, R. M. Alani, E. S. Antonarakis, N. S. Azad, A. Bardelli, H. Brem, J. L. Cameron, C. C. Lee, L. A. Fecher, G. L. Gallia, P. Gibbs, D. Le, R. L. Giuntoli, M. Goggins, M. D. Hogarty, M. Holdhoff, S. M. Hong, Y. Jiao, H. H. Juhl, J. J. Kim, G. Siravegna, D. A. Laheru, C. Lauricella, M. Lim, E. J. Lipson, S. K. Marie, G. J. Netto, K. S. Oliner, A. Olivi, L. Olsson, G. J. Riggins, A. Sartore-Bianchi, K. Schmidt, M. Shih l, S. M. Oba-Shinjo, S. Siena, D. Theodorescu, J. Tie, T. T. Harkins, S. Veronese, T. L. Wang, J. D. Weingart, C. L. Wolfgang, L. D. Wood, D. Xing, R. H. Hruban, J. Wu, P. J. Allen, C. M. Schmidt, M. A. Choti, V. E. Velculescu, K. W. Kinzler, B. Vogelstein, N. Papadopoulos, and L. A. Diaz, Jr. (2014). Detection of circulating tumor DNA in early- and late-stage human malignancies. *Sci Transl Med.* **6**(224): 224ra224.

Bianchi, D. W., R. L. Parker, J. Wentworth, R. Madankumar, C. Saffer, A. F. Das, J. A. Craig, D. I. Chudova, P. L. Devers, K. W. Jones, K. Oliver, R. P. Rava, A. J. Sehnert, and C. S. Group (2014). DNA sequencing versus standard prenatal aneuploidy screening. *N Engl J Med.* **370**(9): 799–808.

Bianchi, D. W., L. D. Platt, J. D. Goldberg, A. Z. Abuhamad, A. J. Sehnert, R. P. Rava, and B. I. S. S. t. A. d. f. a. S. G. MatErnal (2012). Genome-wide fetal aneuploidy detection by maternal plasma DNA sequencing. *Obstet Gynecol.* **119**(5): 890–901.

Botezatu, I., O. Serdyuk, G. Potapova, V. Shelepov, R. Alechina, Y. Molyaka, V. Ananev, I. Bazin, A. Garin, M. Narimanov, V. Knysh, H. Melkonyan, S. Umansky, and A. Lichtenstein (2000). Genetic analysis of DNA excreted in urine: a new approach for detecting specific genomic DNA sequences from cells dying in an organism. *Clin Chem.* **46**(8 Pt 1): 1078–1084.

Carreira, S., A. Romanel, J. Goodall, E. Grist, R. Ferraldeschi, S. Miranda, D. Prandi, D. Lorente, J. S. Frenel, C. Pezaro, A. Omlin, D. N. Rodrigues, P. Flohr, N. Tunariu, S. d. B. J, F. Demichelis, and G. Attard (2014). Tumor clone dynamics in lethal prostate cancer. *Sci Transl Med.* **6**(254): 254ra125.

Chan, R. W., P. Jiang, X. Peng, L. S. Tam, G. J. Liao, E. K. Li, P. C. Wong, H. Sun, K. C. Chan, R. W. Chiu, and Y. M. Lo (2014). Plasma DNA aberrations in systemic lupus erythematosus revealed by genomic and methylomic sequencing. *Proc Natl Acad Sci U S A.* **111**(49): E5302–5311.

Chan, K. C., P. Jiang, Y. W. Zheng, G. J. Liao, H. Sun, J. Wong, S. S. Siu, W. C. Chan, S. L. Chan, A. T. Chan, P. B. Lai, R. W. Chiu, and Y. M. Lo (2013). Cancer genome scanning in plasma: detection of tumor-associated copy number aberrations, single-nucleotide variants, and tumoral heterogeneity by massively parallel sequencing. *Clin Chem.* **59**(1): 211–224.

Chan, K. C. A., J. Zhang, A. B. Hui, N. Wong, T. K. Lau, T. N. Leung, K. W. Lo, D. W. Huang, and Y. M. D. Lo (2004). Size distributions of maternal and fetal DNA in maternal plasma. *Clin Chem.* **50**(1): 88–92.

Chen, E. Z., R. W. Chiu, H. Sun, R. Akolekar, K. C. Chan, T. Y. Leung, P. Jiang, Y. W. Zheng, F. M. Lun, L. Y. Chan, Y. Jin, A. T. Go, E. T. Lau, W. W. To, W. C. Leung, R. Y. Tang, S. K. Au-Yeung, H. Lam, Y. Y. Kung, X. Zhang, J. M. van Vugt, R. Minekawa, M. H. Tang, J. Wang, C. B. Oudejans, T. K. Lau, K. H. Nicolaides, and Y. M. Lo (2011). Noninvasive prenatal diagnosis of fetal trisomy 18 and trisomy 13 by maternal plasma DNA sequencing. *PLoS One.* **6**(7): e21791.

Chim, S. S. C., S. Jin, T. Y. Lee, F. M. F. Lun, W. S. Lee, L. Y. Chan, Y. Jin, N. Yang, Y. K. Tong, T. Y. Leung, T. K. Lau, C. Ding, R. W. K. Chiu, and Y. M. D. Lo (2008). Systematic search for placental DNA-methylation markers on chromosome 21: toward a maternal plasma-based epigenetic test for fetal trisomy 21. *Clin Chem.* **54**(3): 500–511.

Chiu, R. W., R. Akolekar, Y. W. Zheng, T. Y. Leung, H. Sun, K. C. Chan, F. M. Lun, A. T. Go, E. T. Lau, W. W. To, W. C. Leung, R. Y. Tang, S. K. Au-Yeung, H. Lam, Y. Y. Kung, X. Zhang, J. M. van Vugt, R. Minekawa, M. H. Tang, J. Wang, C. B. Oudejans, T. K. Lau, K. H. Nicolaides, and Y. M. Lo (2011). Non-invasive prenatal assessment of trisomy 21 by multiplexed maternal plasma DNA sequencing: large scale validity study. *BMJ.* **342**: c7401.

Chiu, R. W., K. C. Chan, Y. Gao, V. Y. Lau, W. Zheng, T. Y. Leung, C. H. Foo, B. Xie, N. B. Tsui, F. M. Lun, B. C. Zee, T. K. Lau, C. R. Cantor, and Y. M. Lo (2008). Noninvasive prenatal diagnosis of fetal chromosomal aneuploidy by massively parallel genomic sequencing of DNA in maternal plasma. *Proc Natl Acad Sci U S A.* **105**(51): 20458–20463.

Chiu, R. W., L. L. Poon, T. K. Lau, T. N. Leung, E. M. Wong, and Y. M. Lo (2001). Effects of blood-processing protocols on fetal and total DNA quantification in maternal plasma. *Clin Chem.* **47**(9): 1607–1613.

Chiu, R. W. K., C. R. Cantor, and Y. M. D. Lo (2009). Non-invasive prenatal diagnosis by single molecule counting technologies. *Trends Genet.* **25**(7): 324–331.

Chiu, R. W. K., T. K. Lau, P. T. Cheung, Z. Q. Gong, T. N. Leung, and Y. M. D. Lo (2002). Noninvasive prenatal exclusion of congenital adrenal hyperplasia by maternal plasma analysis: a feasibility study. *Clin Chem.* **48**(5): 778–780.

Chiu, R. W. K., T. K. Lau, T. N. Leung, K. C. K. Chow, D. H. Chui, and Y. M. D. Lo (2002). Prenatal exclusion of beta thalassaemia major by examination of maternal plasma. *Lancet.* **360**(9338): 998–1000.

Chiu, R. W. K., H. Sun, R. Akolekar, C. Clouser, C. Lee, K. McKernan, D. Zhou, K. H. Nicolaides, and Y. M. D. Lo (2010). Maternal plasma DNA analysis with massively parallel sequencing by ligation for noninvasive prenatal diagnosis of trisomy 21. *Clin Chem.* **56**(3): 459–463.

Costa, J. M., A. Benachi, E. Gautier, J. M. Jouannic, P. Ernault, and Y. Dumez (2001). First-trimester fetal sex determination in maternal serum using real-time PCR. *Prenat Diagn.* **21**(12): 1070–1074.

Dawson, S. J., D. W. Tsui, M. Murtaza, H. Biggs, O. M. Rueda, S. F. Chin, M. J. Dunning, D. Gale, T. Forshew, B. Mahler-Araujo, S. Rajan, S. Humphray, J. Becq, D. Halsall, M. Wallis, D. Bentley, C. Caldas, and N. Rosenfeld (2013). Analysis of circulating tumor DNA to monitor metastatic breast cancer. *N Engl J Med.* **368**(13): 1199–1209.

De Vlaminck, I., H. A. Valantine, T. M. Snyder, C. Strehl, G. Cohen, H. Luikart, N. F. Neff, J. Okamoto, D. Bernstein, D. Weisshaar, S. R. Quake, and K. K. Khush (2014). Circulating cell-free DNA enables noninvasive diagnosis of heart transplant rejection. *Sci Transl Med.* **6**(241): 241ra277.

Devaney, S. A., G. E. Palomaki, J. A. Scott, and D. W. Bianchi (2011). Noninvasive fetal sex determination using cell-free fetal DNA: a systematic review and meta-analysis. *JAMA.* **306**(6): 627–636.

Diaz, L. A., Jr., R. T. Williams, J. Wu, I. Kinde, J. R. Hecht, J. Berlin, B. Allen, I. Bozic, J. G. Reiter, M. A. Nowak, K. W. Kinzler, K. S. Oliner and B. Vogelstein (2012). The molecular evolution of acquired resistance to targeted EGFR blockade in colorectal cancers. *Nature.* **486**(7404): 537–540.

Diehl, F., K. Schmidt, M. A. Choti, K. Romans, S. Goodman, M. Li, K. Thornton, N. Agrawal, L. Sokoll, S. A. Szabo, K. W. Kinzler, B. Vogelstein and L. A. Diaz, Jr. (2008). Circulating mutant DNA to assess tumor dynamics. *Nat Med.* **14**(9): 985–990.

Ding, C., R. W. K. Chiu, T. K. Lau, T. N. Leung, L. C. Chan, A. Y. Chan, P. Charoenkwan, I. S. Ng, H. Y. Law, E. S. Ma, X. Xu, C. Wanapirak, T. Sanguansermsri, C. Liao, M. A. Ai, D. H. Chui, C. R. Cantor, and Y. M. D. Lo (2004). MS analysis of single-nucleotide differences in circulating nucleic acids: Application to noninvasive prenatal diagnosis. *Proc Natl Acad Sci U S A.* **101**(29): 10762–10767.

Evans, M. I., and M. Kilpatrick (2010). Noninvasive prenatal diagnosis: 2010. *Clin Lab Med.* **30**(3): 655–665.

Fan, H. C., Y. J. Blumenfeld, U. Chitkara, L. Hudgins, and S. R. Quake (2008). Noninvasive diagnosis of fetal aneuploidy by shotgun sequencing DNA from maternal blood. *Proc Natl Acad Sci U S A.* **105**(42): 16266–16271.

Fan, H. C., Y. J. Blumenfeld, U. Chitkara, L. Hudgins, and S. R. Quake (2010). Analysis of the size distributions of fetal and maternal cell-free DNA by paired-end sequencing. *Clinical chemistry.* **56**(8): 1279–1286.

Fan, H. C., W. Gu, J. Wang, Y. J. Blumenfeld, Y. Y. El-Sayed, and S. R. Quake (2012). Non-invasive prenatal measurement of the fetal genome. *Nature.* **487**(7407): 320–324.

Finning, K., P. Martin, J. Summers, E. Massey, G. Poole, and G. Daniels (2008). Effect of high throughput RHD typing of fetal DNA in maternal plasma on use of anti-RhD immunoglobulin in RhD negative pregnant women: prospective feasibility study. *BMJ.* **336**(7648): 816–818.

Forshew, T., M. Murtaza, C. Parkinson, D. Gale, D. W. Tsui, F. Kaper, S. J. Dawson, A. M. Piskorz, M. Jimenez-Linan, D. Bentley, J. Hadfield, A. P. May, C. Caldas, J. D. Brenton, and N. Rosenfeld (2012). Noninvasive identification and monitoring of cancer mutations by targeted deep sequencing of plasma DNA. *Sci Transl Med.* **4**(136): 136ra168.

Gerlinger, M., A. J. Rowan, S. Horswell, J. Larkin, D. Endesfelder, E. Gronroos, P. Martinez, N. Matthews, A. Stewart, P. Tarpey, I. Varela, B. Phillimore, S. Begum, N. Q. McDonald, A. Butler, D. Jones, K. Raine, C. Latimer, C. R. Santos, M. Nohadani, A. C. Eklund, B. Spencer-Dene, G. Clark, L. Pickering, G. Stamp, M. Gore, Z. Szallasi, J. Downward, P. A. Futreal, and C. Swanton (2012). Intratumor heterogeneity and branched evolution revealed by multiregion sequencing. *N Engl J Med.* **366**(10): 883–892.

Gerstung, M., C. Beisel, M. Rechsteiner, P. Wild, P. Schraml, H. Moch, and N. Beerenwinkel (2012). Reliable detection of subclonal single-nucleotide variants in tumour cell populations. *Nat Commun.* **3**: 811.

Go, A. T., A. Visser, M. A. Mulders, M. A. Blankenstein, J. M. van Vugt, and C. B. Oudejans (2007). 44 single-nucleotide polymorphisms expressed by placental RNA: assessment for use in noninvasive prenatal diagnosis of trisomy 21. *Clin Chem.* **53**(12): 2223–2224.

Gut, I. G. (2004). DNA analysis by MALDI-TOF mass spectrometry. *Hum Mutat.* **23**(5): 437–441.

Hill, M., C. Lewis, L. Jenkins, S. Allen, R. G. Elles, and L. S. Chitty (2012). Implementing noninvasive prenatal fetal sex determination using cell-free fetal DNA in the United Kingdom. *Expert Opin Biol Ther.* **12 Suppl 1**: S119–126.

Holdenrieder, S., P. Stieber, L. Y. Chan, S. Geiger, A. Kremer, D. Nagel, and Y. M. Lo (2005). Cell-free DNA in serum and plasma: comparison of ELISA and quantitative PCR. *Clin Chem.* **51**(8): 1544–1546.

Honda, H., N. Miharu, Y. Ohashi, O. Samura, M. Kinutani, T. Hara, and K. Ohama (2002). Fetal gender determination in early pregnancy through qualitative, and quantitative analysis of fetal DNA in maternal serum. *Hum Genet.* **110**(1): 75–79.

Hung, E. C., T. K. Shing, S. S. Chim, P. C. Yeung, R. W. Chan, K. W. Chik, V. Lee, N. B. Tsui, C. K. Li, C. S. Wong, R. W. Chiu, and Y. M. Lo (2009). Presence of donor-derived DNA, and cells in the urine of sex-mismatched hematopoietic stem cell transplant recipients: implication for the transrenal hypothesis. *Clin Chem.* **55**(4): 715–722.

Hyman, D. M., E. L. Diamond, C. R. Vibat, L. Hassaine, J. C. Poole, M. Patel, V. R. Holley, G. Cabrilo, T. T. Lu, M. E. Arcila, Y. R. Chung, R. Rampal, M. E. Lacouture, N. Rosen, F. Meric-Bernstam, J. Baselga, R. Kurzrock, M. G. Erlander, F. Janku, and O. Abdel-Wahab (2015). Prospective blinded study of BRAFV600E mutation detection in cell-free DNA of patients with systemic histiocytic disorders. *Cancer Discov.* **5**(1): 64–71.

Jahr, S., H. Hentze, S. Englisch, D. Hardt, F. O. Fackelmayer, R. D. Hesch, and R. Knippers (2001). DNA fragments in the blood plasma of cancer patients: quantitations and evidence for their origin from apoptotic and necrotic cells. *Cancer Res.* **61**(4): 1659–1665.

Jiang, P., C. W. Chan, K. C. Chan, S. H. Cheng, J. Wong, V. W. Wong, G. L. Wong, S. L. Chan, T. S. Mok, H. L. Chan, P. B. Lai, R. W. Chiu, and Y. M. Lo (2015). Lengthening and shortening of plasma DNA in hepatocellular carcinoma patients. *Proc Natl Acad Sci U S A.*

Kinde, I., J. Wu, N. Papadopoulos, K. W. Kinzler, and B. Vogelstein (2011). Detection and quantification of rare mutations with massively parallel sequencing. *Proc Natl Acad Sci U S A.* **108**(23): 9530–9535.

Kitzman, J. O., M. W. Snyder, M. Ventura, A. P. Lewis, R. Qiu, L. E. Simmons, H. S. Gammill, C. E. Rubens, D. A. Santillan, J. C. Murray, H. K. Tabor, M. J. Bamshad, E. E. Eichler, and J. Shendure (2012). Noninvasive whole-genome sequencing of a human fetus. *Sci Transl Med.* **4**(137): 137ra176.

Koffler, D., V. Agnello, R. Winchester, and H. G. Kunkel (1973). The occurrence of single-stranded DNA in the serum of patients with systemic lupus erythematosus and other diseases. *J Clin Invest.* **52**(1): 198–204.

Leary, R. J., M. Sausen, I. Kinde, N. Papadopoulos, J. D. Carpten, D. Craig, J. O'Shaughnessy, K. W. Kinzler, G. Parmigiani, B. Vogelstein, L. A. Diaz, Jr., and V. E. Velculescu (2012). Detection of chromosomal alterations in the circulation of cancer patients with whole-genome sequencing. *Sci Transl Med.* **4**(162): 162ra154.

Lebofsky, R., C. Decraene, V. Bernard, M. Kamal, A. Blin, Q. Leroy, T. Rio Frio, G. Pierron, C. Callens, I. Bieche, A. Saliou, J. Madic, E. Rouleau, F. C. Bidard, O. Lantz, M. H. Stern, C. Le Tourneau, and J. Y. Pierga (2014). Circulating tumor DNA as a non-invasive substitute to metastasis biopsy for tumor genotyping and personalized medicine in a prospective trial across all tumor types. *Mol Oncol.*

Leon, S. A., B. Shapiro, D. M. Sklaroff, and M. J. Yaros (1977). Free DNA in the serum of cancer patients and the effect of therapy. *Cancer Res.* **37**(3): 646–650.

Li, J., L. Wang, H. Mamon, M. H. Kulke, R. Berbeco, and G. M. Makrigiorgos (2008). Replacing PCR with COLD-PCR enriches variant DNA sequences and redefines the sensitivity of genetic testing. *Nat Med.* **14**(5): 579–584.

Lipson, E. J., V. E. Velculescu, T. S. Pritchard, M. Sausen, D. M. Pardoll, S. L. Topalian, and L. A. Diaz, Jr. (2014). Circulating tumor DNA analysis as a real-time method for monitoring tumor burden in melanoma patients undergoing treatment with immune checkpoint blockade. *J Immunother Cancer.* **2**(1): 42.

Lo, Y. M., K. C. Chan, H. Sun, E. Z. Chen, P. Jiang, F. M. Lun, Y. W. Zheng, T. Y. Leung, T. K. Lau, C. R. Cantor, and R. W. Chiu (2010). Maternal plasma DNA sequencing reveals the genome-wide genetic and mutational profile of the fetus. *Sci Transl Med.* **2**(61): 61ra91.

Lo, Y. M., and R. W. Chiu (2007). Prenatal diagnosis: progress through plasma nucleic acids. *Nat Rev Genet.* **8**(1): 71–77.

Lo, Y. M., and R. W. Chiu (2012). Genomic analysis of fetal nucleic acids in maternal blood. *Annu Rev Genomics Hum Genet.* **13**: 285–306.

Lo, Y. M., T. H. Rainer, L. Y. Chan, N. M. Hjelm, and R. A. Cocks (2000). Plasma DNA as a prognostic marker in trauma patients. *Clin Chem.* **46**(3): 319–323.

Lo, Y. M. D., N. Corbetta, P. F. Chamberlain, V. Rai, I. L. Sargent, C. W. Redman, and J. S. Wainscoat (1997). Presence of fetal DNA in maternal plasma and serum. *Lancet.* **350**(9076): 485–487.

Lo, Y. M. D., N. M. Hjelm, C. Fidler, I. L. Sargent, M. F. Murphy, P. F. Chamberlain, P. M. Poon, C. W. Redman, and J. S. Wainscoat (1998). Prenatal diagnosis of fetal RhD status by molecular analysis of maternal plasma. *N Engl J Med.* **339**(24): 1734–1738.

Lo, Y. M. D., T. N. Leung, M. S. Tein, I. L. Sargent, J. Zhang, T. K. Lau, C. J. Haines, and C. W. Redman (1999). Quantitative abnormalities of fetal DNA in maternal serum in preeclampsia. *Clin Chem.* **45**(2): 184–188.

Lo, Y. M. D., N. B. Y. Tsui, R. W. K. Chiu, T. K. Lau, T. N. Leung, M. M. Heung, A. Gerovassili, Y. Jin, K. H. Nicolaides, C. R. Cantor, and C. Ding (2007). Plasma placental RNA allelic ratio permits noninvasive prenatal chromosomal aneuploidy detection. *Nat Med.* **13**(2): 218–223.

Lui, Y. Y., K. W. Chik, R. W. Chiu, C. Y. Ho, C. W. Lam, and Y. M. Lo (2002). Predominant hematopoietic origin of cell-free DNA in plasma and serum after sex-mismatched bone marrow transplantation. *Clin Chem.* **48**(3): 421–427.

Lui, Y. Y., and Y. M. Dennis (2002). Circulating DNA in plasma and serum: biology, preanalytical issues and diagnostic applications. *Clin Chem Lab Med.* **40**(10): 962–968.

McBride, D. J., A. K. Orpana, C. Sotiriou, H. Joensuu, P. J. Stephens, L. J. Mudie, E. Hamalainen, L. A. Stebbings, L. C. Andersson, A. M. Flanagan, V. Durbecq, M. Ignatiadis, O. Kallioniemi, C. A. Heckman, K. Alitalo, H. Edgren, P. A. Futreal, M. R. Stratton, and P. J. Campbell (2010). Use of cancer-specific genomic rearrangements to quantify disease burden in plasma from patients with solid tumors. *Genes Chromosomes Cancer.* **49**(11): 1062–1069.

Mendel, P., L. Mandel, and et al. (1948). Les acides nucleiques du plasma sanguin chez l'homme. *C R Hebd Seances Acad Sci.* **226**(24): 2019–2021.

Misale, S., R. Yaeger, S. Hobor, E. Scala, M. Janakiraman, D. Liska, E. Valtorta, R. Schiavo, M. Buscarino, G. Siravegna, K. Bencardino, A. Cercek, C. T. Chen, S. Veronese, C. Zanon, A. Sartore-Bianchi, M. Gambacorta, M. Gallicchio, E. Vakiani, V. Boscaro, E. Medico, M. Weiser, S. Siena, F. Di Nicolantonio, D. Solit, and A. Bardelli (2012). Emergence of KRAS mutations and acquired resistance to anti-EGFR therapy in colorectal cancer. *Nature.* **486**(7404): 532–536.

Mouliere, F., S. El Messaoudi, C. Gongora, A. S. Guedj, B. Robert, M. Del Rio, F. Molina, P. J. Lamy, E. Lopez-Crapez, M. Mathonnet, M. Ychou, D. Pezet, and A. R. Thierry (2013). Circulating Cell-Free DNA from Colorectal Cancer Patients May Reveal High KRAS or BRAF Mutation Load. *Transl Oncol.* **6**(3): 319–328.

Murtaza, M., S. J. Dawson, D. W. Tsui, D. Gale, T. Forshew, A. M. Piskorz, C. Parkinson, S. F. Chin, Z. Kingsbury, A. S. Wong, F. Marass, S. Humphray, J. Hadfield, D. Bentley, T. M. Chin, J. D. Brenton, C. Caldas, and N. Rosenfeld (2013). Non-invasive analysis of acquired resistance to cancer therapy by sequencing of plasma DNA. *Nature.* **497**(7447): 108–112.

Newman, A. M., S. V. Bratman, J. To, J. F. Wynne, N. C. Eclov, L. A. Modlin, C. L. Liu, J. W. Neal, H. A. Wakelee, R. E. Merritt, J. B. Shrager, B. W. Loo, Jr., A. A. Alizadeh, and M. Diehn (2014). An ultrasensitive method for quantitating circulating tumor DNA with broad patient coverage. *Nat Med.* **20**(5): 548–554.

Ng, E. K., T. N. Leung, N. B. Y. Tsui, T. K. Lau, N. S. Panesar, R. W. K. Chiu, and Y. M. D. Lo (2003). The concentration of circulating corticotropin-releasing hormone mRNA in maternal plasma is increased in preeclampsia. *Clin Chem.* **49**(5): 727–731.

Nygren, A. O., J. Dean, T. J. Jensen, S. Kruse, W. Kwong, D. van den Boom and M. Ehrich (2010). Quantification of fetal DNA by use of methylation-based DNA discrimination. *Clin Chem.* **56**(10): 1627–1635.

Oxnard, G. R., C. P. Paweletz, Y. Kuang, S. L. Mach, A. O'Connell, M. M. Messineo, J. J. Luke, M. Butaney, P. Kirschmeier, D. M. Jackman, and P. A. Janne (2014). Noninvasive detection of response and resistance in EGFR-mutant lung cancer using quantitative next-generation genotyping of cell-free plasma DNA. *Clin Cancer Res.* **20**(6): 1698–1705.

Palomaki, G. E., E. M. Kloza, G. M. Lambert-Messerlian, J. E. Haddow, L. M. Neveux, M. Ehrich, D. van den Boom, A. T. Bombard, C. Deciu, W. W. Grody, S. F. Nelson, and J. A. Canick (2011). DNA sequencing of maternal plasma to detect Down syndrome: an international clinical validation study. *Genet Med.* **13**(11): 913–920.

Parker, S. E., C. T. Mai, M. A. Canfield, R. Rickard, Y. Wang, R. E. Meyer, P. Anderson, C. A. Mason, J. S. Collins, R. S. Kirby, A. Correa, and N. National Birth Defects Prevention (2010). Updated National Birth Prevalence estimates for selected birth defects in the United States, 2004–2006. *Birth Defects Res A Clin Mol Teratol.* **88**(12): 1008–1016.

Pinheiro, L. B., V. A. Coleman, C. M. Hindson, J. Herrmann, B. J. Hindson, S. Bhat, and K. R. Emslie (2012). Evaluation of a droplet digital polymerase chain reaction format for DNA copy number quantification. *Anal Chem.* **84**(2): 1003–1011.

Rainer, T. H., Y. M. Lo, L. Y. Chan, N. Y. Lam, L. C. Lit, and R. A. Cocks (2001). Derivation of a prediction rule for posttraumatic organ failure using plasma DNA and other variables. *Ann N Y Acad Sci.* **945**: 211–220.

Raptis, L., and H. A. Menard (1980). Quantitation and characterization of plasma DNA in normals and patients with systemic lupus erythematosus. *J Clin Invest.* **66**(6): 1391–1399.

Rhodes, A., S. J. Wort, H. Thomas, P. Collinson, and E. D. Bennett (2006). Plasma DNA concentration as a predictor of mortality and sepsis in critically ill patients. *Crit Care.* **10**(2): R60.

Saukkonen, K., P. Lakkisto, V. Pettila, M. Varpula, S. Karlsson, E. Ruokonen, K. Pulkki, and G. Finnsepsis Study (2008). Cell-free plasma DNA as a predictor of outcome in severe sepsis and septic shock. *Clin Chem.* **54**(6): 1000–1007.

Scheffer, P. G., C. E. van der Schoot, G. C. Page-Christiaens, and M. de Haas (2011). Noninvasive fetal blood group genotyping of rhesus D, c, E and of K in alloimmunised pregnant women: evaluation of a 7-year clinical experience. *BJOG.* **118**(11): 1340–1348.

Sekizawa, A., M. Jimbo, H. Saito, M. Iwasaki, Y. Sugito, Y. Yukimoto, J. Otsuka, and T. Okai (2002). Increased cell-free fetal DNA in plasma of two women with invasive placenta. *Clin Chem.* **48**(2): 353–354.

Shi, H., W. Hugo, X. Kong, A. Hong, R. C. Koya, G. Moriceau, T. Chodon, R. Guo, D. B. Johnson, K. B. Dahlman, M. C. Kelley, R. F. Kefford, B. Chmielowski, J. A. Glaspy, J. A. Sosman, N. van Baren, G. V. Long, A. Ribas and R. S. Lo (2014). Acquired resistance and clonal evolution in melanoma during BRAF inhibitor therapy. *Cancer Discov.* **4**(1): 80–93.

Snyder, T. M., K. K. Khush, H. A. Valantine, and S. R. Quake (2011). Universal noninvasive detection of solid organ transplant rejection. *Proc Natl Acad Sci U S A.* **108**(15): 6229–6234.

Stroun, M., P. Anker, P. Maurice, and P. B. Gahan (1977). Circulating nucleic acids in higher organisms. *Int Rev Cytol.* **51**: 1–48.

Sugito, Y., A. Sekizawa, A. Farina, Y. Yukimoto, H. Saito, M. Iwasaki, N. Rizzo, and T. Okai (2003). Relationship between severity of hyperemesis gravidarum and fetal DNA concentration in maternal plasma. *Clin Chem.* **49**(10): 1667–1669.

Taly, V., D. Pekin, L. Benhaim, S. K. Kotsopoulos, D. Le Corre, X. Li, I. Atochin, D. R. Link, A. D. Griffiths, K. Pallier, H. Blons, O. Bouche, B. Landi, J. B. Hutchison, and P. Laurent-Puig (2013). Multiplex picodroplet digital PCR to detect KRAS mutations in circulating DNA from the plasma of colorectal cancer patients. *Clin Chem.* **59**(12): 1722–1731.

Tan, E. M., P. H. Schur, R. I. Carr, and H. G. Kunkel (1966). Deoxybonucleic acid (DNA) and antibodies to DNA in the serum of patients with systemic lupus erythematosus. *J Clin Invest.* **45**(11): 1732–1740.

Thierry, A. R., F. Mouliere, S. El Messaoudi, C. Mollevi, E. Lopez-Crapez, F. Rolet, B. Gillet, C. Gongora, P. Dechelotte, B. Robert, M. Del Rio, P. J. Lamy, F. Bibeau, M. Nouaille, V. Loriot, A. S. Jarrousse, F. Molina, M. Mathonnet, D. Pezet, and M. Ychou (2014). Clinical validation of the detection of KRAS and BRAF mutations from circulating tumor DNA. *Nat Med.* **20**(4): 430–435.

Thompson, J. D., G. Shibahara, S. Rajan, J. Pel, and A. Marziali (2012). Winnowing DNA for rare sequences: highly specific sequence and methylation based enrichment. *PLoS One*. **7**(2): e31597.

Tsui, N. B., R. A. Kadir, K. C. Chan, C. Chi, G. Mellars, E. G. Tuddenham, T. Y. Leung, T. K. Lau, R. W. Chiu, and Y. M. Lo (2011). Noninvasive prenatal diagnosis of hemophilia by microfluidics digital PCR analysis of maternal plasma DNA. *Blood*. **117**(13): 3684–3691.

Vogelstein, B., and K. W. Kinzler (1999). Digital PCR. *Proc Natl Acad Sci U S A*. **96**(16): 9236–9241.

Wong, D., S. Moturi, V. Angkachatchai, R. Mueller, G. DeSantis, D. van den Boom, and M. Ehrich (2013). Optimizing blood collection, transport and storage conditions for cell free DNA increases access to prenatal testing. *Clin Biochem*. **46**(12): 1099–1104.

Yamanouchi, S., D. Kudo, M. Yamada, N. Miyagawa, H. Furukawa, and S. Kushimoto (2013). Plasma mitochondrial DNA levels in patients with trauma and severe sepsis: time course and the association with clinical status. *J Crit Care*. **28**(6): 1027–1031.

Yu, S. C., K. C. Chan, Y. W. Zheng, P. Jiang, G. J. Liao, H. Sun, R. Akolekar, T. Y. Leung, A. T. Go, J. M. van Vugt, R. Minekawa, C. B. Oudejans, K. H. Nicolaides, R. W. Chiu, and Y. M. Lo (2014). Size-based molecular diagnostics using plasma DNA for noninvasive prenatal testing. *Proc Natl Acad Sci U S A*. **111**(23): 8583–8588.

Yu, S. C., S. W. Lee, P. Jiang, T. Y. Leung, K. C. Chan, R. W. Chiu, and Y. M. Lo (2013). High-resolution profiling of fetal DNA clearance from maternal plasma by massively parallel sequencing. *Clin Chem*. **59**(8): 1228–1237.

Yung, T. K., K. C. Chan, T. S. Mok, J. Tong, K. F. To, and Y. M. Lo (2009). Single-molecule detection of epidermal growth factor receptor mutations in plasma by microfluidics digital PCR in non-small cell lung cancer patients. *Clin Cancer Res*. **15**(6): 2076–2084.

Chapter 5

Circulating Tumor Cells and Personalized Medicine

Jin Sun Lee, Mark Jesus M. Magbanua, Marc R. Jabon, and John W. Park

Division of Hematology Oncology, Helen Diller Comprehensive Cancer Center, University of California San Francisco, San Francisco, California, USA

mark.magbanua@ucsf.edu

Extremely rare circulating tumor cells (CTCs) in blood can now be reliably detected and enumerated. Numerous studies have shown that elevated levels of CTCs in cancer patients are associated with poor prognosis. Furthermore, recent efforts towards molecular characterization of CTCs have provided novel insights into the biology of these tumor cells. In this chapter, we describe methods for detection and enumeration of CTCs. We also outline the clinical implications of CTC enumeration and molecular profiling and discuss clinical trials that have utilized information from CTCs in the context of personalized medicine.

5.1 Metastasis and Circulating Tumor Cells

Metastasis is the major cause of death among cancer patients. Despite its vast clinical importance, the molecular and cellular

---

Cancer Genetics and Genomics for Personalized Medicine
Edited by Il-Jin Kim

ISBN 978-981-4669-87-0 (Hardcover), 978-1-315-36476-6 (eBook)
www.panstanford.com

mechanisms involved in the metastatic cascade are poorly understood. The paucity of knowledge can be attributed to the difficulty in obtaining metastatic tissue, which often requires invasive procedures. Tumor cells detected in the blood, therefore, offer a unique source of metastatic tissue that can be accessed in a relatively non-invasive manner. In the past decade, the field of CTC research has become a very active area of research in oncology. Enumeration studies have unequivocally demonstrated the prognostic significance of these cells. In addition, genomic characterization of CTCs has been actively pursued in an effort to understand the molecular nature of these cells. However, huge technological and technical challenges have been faced in isolating and detecting these extremely rare cells.

5.1.1 The Metastatic Process

The early steps in cancer formation involve the active proliferation of aberrant cells, leading to the formation of a primary tumor. As cancer progresses, tumor cells leave the primary lesion and spread systematically to colonize distant sites. While early-stage cancers (i.e., those that are localized or organ-confined) can potentially be treated, there are no cures available for cancers that have metastasized.

Molecular and cellular determinants that induce metastasis have yet to be fully discovered, but the major steps preceding metastasis (as described below) are widely known.[1,2]

(1) *Local invasion*: Tumor cells in the primary tumor lose intercellular adherence. The cells attach to the basement membrane and release collagenase, which dissolves the basement membrane. As the membrane disintegrates, tumor cells infiltrate the stroma and gain access to the blood vessels and the lymphatics.
(2) *Intravasation*: Tumor cells enter the lumen of blood vessels. Also, as the primary tumor grows, growth factors that stimulate the formation of new blood vessels in a process called angiogenesis are released. Tumor cells can then use these new vessels to enter the circulatory system. The lymphatic system can also serve as another route during intravasation.

(3) *Circulation*: Tumor cells that enter the vessels circulate in the blood. Cells that survive may reach the capillary bed of distant organs.

(4) *Extravasation and implantation*: Tumor cells can then escape from the vasculature and begin colonization of target organs.

(5) *Proliferation and overt metastasis*: Arrested cells adapt to the foreign microenvironment and then initiate proliferation. After successful rounds of proliferation, the metastatic lesions may become clinically detectable.

5.1.2 Circulating Tumor Cells

Tumor cells that enter the circulatory system are referred to as CTCs. These are extremely rare (approximately 1 CTC for every 10^9 blood cells); hence, detecting CTCs can be very technically challenging. Although Thomas Ashworth first described CTCs almost 150 years ago,[3] research on CTCs has only been actively explored in the past few decades. In the 1990s, several groups demonstrated the detection of CTCs using immunologic and molecular techniques.[4–7] A variety of new approaches have since been applied to detect and analyze CTCs (see next section).

Molecular characterization of CTCs offers a unique window into the metastatic process. In addition, there are several advantages of using CTCs as surrogates for monitoring systemic disease. For example, a peripheral blood draw is a much less invasive procedure than a tumor biopsy. Consequently, frequent follow-up is possible with blood-based assays allowing a longitudinal "liquid biopsy" to keep track of changes in tumor characteristics over the course of therapy (Fig. 5.1). In addition, the levels of CTCs in the blood and the molecular information derived from these cells may facilitate early changes in therapy based on real-time results.

5.2 Enrichment and Detection of CTCs

Reliable detection of CTCs requires techniques with exquisite specificity and sensitivity. Most current technologies for CTC detection involve a two-part process composed of an enrichment

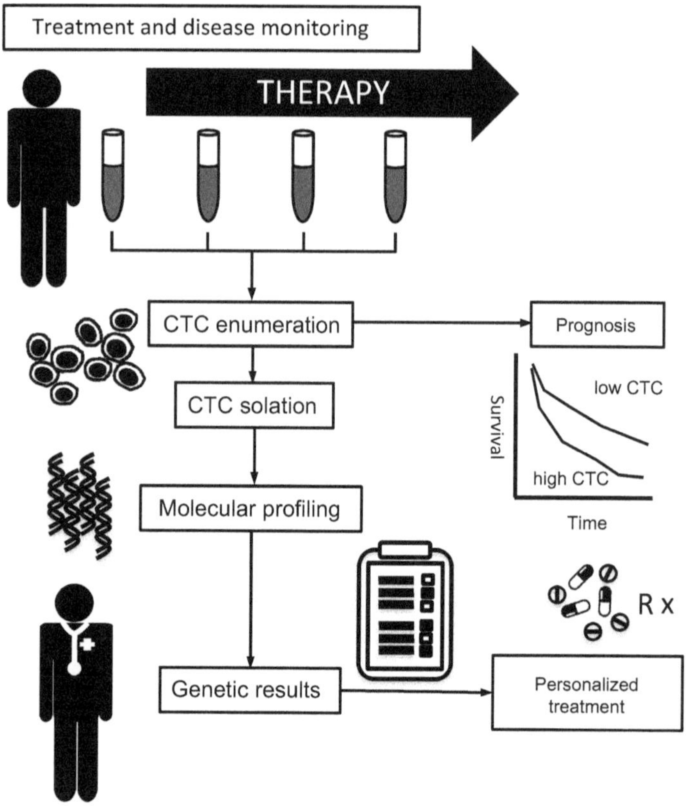

Figure 5.1 CTC analysis and personalized medicine. A schematic diagram showing circulating tumor cell (CTC) analysis in serial blood samples from a cancer patient undergoing therapy. Enumeration of CTCs can be used to determine prognosis while results from genetic profiling can be used to tailor treatment (modified from Lee at al.[105] with permissions).

method to separate CTCs from the bulk of the blood followed by a detection method to distinguish CTCs from an overwhelming number of remaining background blood cells.

5.2.1 Enrichment

Enrichment methods rely on physical characteristics (i.e., size, shape, or density) and/or the biological characteristics (i.e., cell surface protein expression, cell adhesion properties, etc.) of tumor cells.

Immunologic methods, in particular, target cell surface proteins expressed by tumor cells of epithelial origin. For example, the epithelial cell adhesion molecule (EPCAM) is a surface marker widely used for enrichment. Positive selection using magnetic beads coated with anti-EPCAM antibodies (immunomagnetic beads) allows for the separation of CTCs from blood cells. The EPCAM antibodies bind to CTCs, and a magnet is then used to capture cells labeled with beads. It is important to note, however, that epithelial marker-dependent approaches may fail to enrich for tumor cells that have undergone EMT (epithelial-to-mesenchymal transition), which is usually accompanied by down-regulation of epithelial markers.[8] A novel approach that does not utilize specific surface markers is the CAM (collagen adherence matrix) assay (e.g., Vita Cap™).[9] In this method, CTCs adhere to a collagen matrix while blood cells are washed away. It is also possible to use an antibody specific to blood cell markers (e.g., CD45) for negative selection.[10]

Other enrichment methods take advantage of the differences in physical properties (i.e., cell size, shape, or density) between CTCs and blood cells. For example, density-gradient centrifugation facilitates the separation of white blood cells and CTCs from red blood cells (e.g., Ficoll™, OncoQuick®).[11,12] Density gradient-based techniques are less expensive, relatively simple, and do not rely on specific marker expression but have lower enrichment yields and are less specific than immunologic methods.[13] In addition, size exclusion methods (e.g., ISET: isolation by size of epithelial tumor cells) can be used since CTCs are generally larger than erythrocytes and leukocytes. Size-exclusion methods typically utilize special membrane filters. In addition to the size differences between blood cells and CTCs, cell deformability must also be taken into account when using or designing a filtration method to capture CTCs.[14] Like density gradient-based methods, size-exclusion methods do not rely on surface marker expression and are therefore less likely to miss CTCs with down-regulated epithelial marker expression. However, size-based methods may fail to capture CTCs towards the smaller end of the spectrum. Some methods rely on both surface protein expression and physical properties (e.g., CTC-Chip)[15–17] to enrich for CTCs.

5.2.2 Detection

5.2.2.1 Immuno-cytomorphological approach: Immunocytochemistry (ICC)

After the enrichment process, CTCs can be detected via ICC. The CellSearch® (Janssen Diagnostics) system is a semi-automated ICC method used for CTC enumeration. It is currently the only CTC assay that has been cleared by the U.S. Food and Drug Administration for clinical use. The first part of the procedure involves immunomagnetic enrichment using ferrofluid nanoparticles coated with antibodies that target EPCAM. Next, the tumor-enriched sample is stained with fluorophore-conjugated anti-CK and anti-CD45 antibodies to allow for the identification of epithelial cells (i.e., CTCs) and leukocytes, respectively. DAPI (4′, 6-diamidino-2-phenylindole), a DNA stain, is also added to allow for the identification of nucleated cells. The stained CTC-enriched sample is collected into a specially designed cartridge. All of the aforementioned enrichment and staining processes are performed by an automated system (CellTracks® Autoprep®). Next, the cartridge is manually transferred to a semi-automated system (CellTracks® Analyzer® II), which automatically acquires images of tentative CTCs. Lastly, the images of the tentative CTCs are displayed on a computer monitor for final review and enumeration by a trained technician. In this assay, CTCs are defined as nucleated cells that are cytokeratin (CK)-positive, CD45-negative, and at least 4 μm by 4 μm in size (Fig. 5.2). An initial clinical validation study of CellSearch® showed that CTCs are mostly undetectable in non-malignant disease but are present in various metastatic carcinomas.[18] Numerous CTC enumeration studies using CellSearch® have been reported since then (see next section). As mentioned, this EPCAM-dependent method, however, would miss subpopulations of CTCs wherein epithelial marker expression is down-regulated.[19]

More recently, lab-on-a-chip devices (e.g., CTC-Chip) have been developed to miniaturize tools for CTC detection. These novel platforms capture tumor cells in microfluidic chips using different approaches such as size-based separation[20] or by functionalizing surfaces with anti-EPCAM.[15–17] Analogous to the CellSearch® method, CTCs are verified via on-chip ICC staining. In the newest

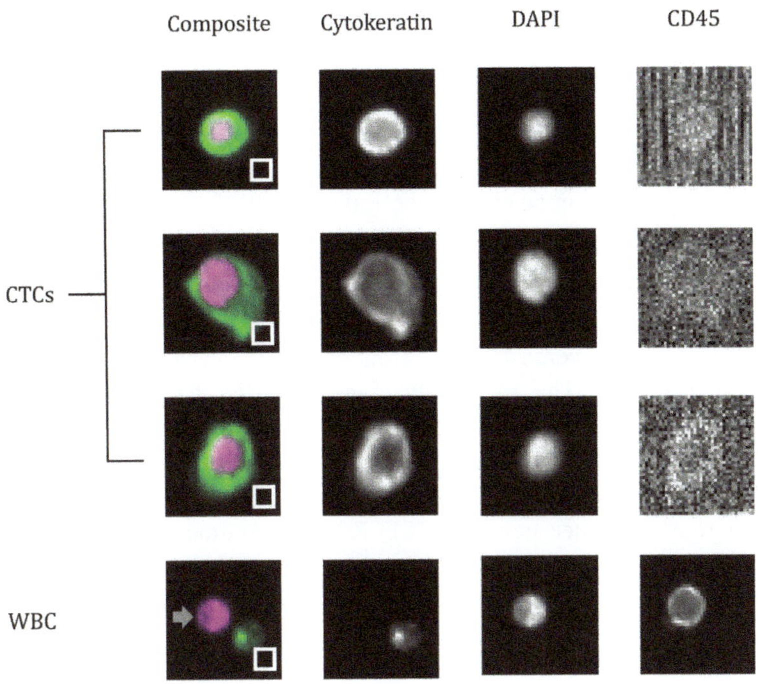

Figure 5.2 CTC detection and enumeration by CellSearch®. Images taken from CellTracks Analyzer II®. CTCs are enumerated by counting cells that are cytokeratin-positive, nucleated (DAPI-positive), CD45-negative, and at least 4 microns by 4 microns (reference square). Contaminating white blood cells (WBC; see arrow) can be identified by the expression of a leukocyte-specific marker CD45 and the lack of cytokeratin expression.

iteration of CTC-Chip, the CTC-iChip, CTCs are collected into a separate container from which they can be detected and enumerated by ICC or characterized using molecular techniques.[21] Whole blood stained with anti-CD45 immunomagnetic beads is gently forced through the chip using an applied pressure. Within the chip, the blood first goes through an array of posts that deflect nucleated cells (e.g., CTCs and white blood cells) and red blood cells in opposite directions (hydrodynamic size-based separation). The CTCs and white blood cells subsequently flow through a channel of a certain geometry that induces the cells to follow a specific trajectory upon exiting the channel (inertial focusing). Finally, the cells pass through a magnetic field in which the immunomagnetically labeled white blood cells deviate from their initial trajectory while the CTCs continue to be captured (magnetophoresis).

5.2.2.2 Molecular approach: Real-time reverse transcription polymerase chain reaction (RT-PCR)

RT-PCR is commonly used to evaluate the presence of CTCs via the detection of CTC-derived mRNA. The process involves RNA extraction from cells in enriched samples, followed by cDNA synthesis by reverse transcription, amplification, and analysis of PCR products. The main advantage of using RT-PCR for CTC detection is that this method is very sensitive. However, the high sensitivity comes at a price since illegitimate transcription of markers in non-malignant cells may lead to false positives. Another drawback of using RT-PCR is that enumeration of intact cells is not possible since CTCs are destroyed in the process and because it is difficult to translate mRNA detection to precise CTC counts.

AdnaTest™ is a commercialized assay for CTC detection that is based on RT-PCR. In this test, immunomagnetic enrichment is performed, and the expression of a panel of associated tumor markers—*HER2/MUC1* for breast cancer and *EGFR/CEA* for colorectal cancer—is assessed via multiplex RT-PCR.[22,23] As with other nucleic acid-based detection methods, AdnaTest™ cannot provide accurate estimates of the numbers of CTCs.

5.3 Clinical Implications of CTC Detection and Enumeration

Early studies in the CTC field were focused on elucidating the clinical relevance of CTCs detected in the blood. The introduction of the CellSearch® system in 2004 allowed for the standardization of CTC enumeration and further accelerated interest in this research area.[18] Initial CellSearch® studies in patients with metastatic breast cancer demonstrated that elevated numbers of CTCs were associated with impaired clinical outcomes. Subsequent studies using CellSearch® as well as RT-PCR-based detection of CTCs in breast and other solid tumors further confirmed the prognostic value of these cells.

5.3.1 Breast Cancer

The most active field of research investigating the clinical value of CTCs has been in the metastatic breast cancer setting. Results from several clinical trials using the CellSearch® system have shown that increased levels of CTCs or the failure to reduce the number of CTCs during therapy are associated with poor outcomes.[24–31] In 2004, Cristofanilli et al. reported that the presence of ≥5 CTCs in 7.5 mL of peripheral blood was associated with reduced survival.[32] This study was the first to demonstrate that the level of CTCs before treatment is an independent prognostic factor in metastatic breast cancer. Serial analysis of CTCs by Hayes et al. showed that increased CTC numbers at any time point was a reliable indicator of impaired survival.[33] Recent meta-analysis conducted in 1944 patients from 20 studies confirmed the independent prognostic effect of CTC counts on progression-free survival (PFS) and overall survival (OS).[34] Recently, the results of the SWOG (Southwest Oncology Group) S0500 trial demonstrated that early switching to an alternative chemotherapy regimen in patients with persistently high CTCs did not result in better survival.[35]

CTC enumeration studies using the CellSearch® system have also been performed in non-metastatic breast cancer settings. Lucci et al. found that the presence of CTCs in ≥7.5 mL in the blood of early-stage patients predicts impaired clinical outcomes as well as early recurrence.[36] Bidard et al. also reported that detection of at least one CTC in 7.5 mL of blood before neoadjuvant chemotherapy was significantly associated with poor survival.[37] The SUCCESS study has provided strong evidence for the prognostic relevance of CTCs from a large prospective trial of 2026 patients before adjuvant chemotherapy and 1492 patients after chemotherapy.[38] In this study, the presence of at least one CTC per 30 mL of blood before or after adjuvant chemotherapy was an independent prognostic factor, and elevated CTC counts were associated with poor survival.

Taken together, these studies highlight the potential clinical utility of CTCs as a prognostic tool in early and advanced stage breast cancers but also underscore the need for further studies to understand how CTC numbers can be used for individualized therapy with the goal of improving clinical outcomes.

5.3.2 Prostate Cancer

The metastatic prostate cancer setting has been another active field for CTC research. Danila et al. enumerated CTCs from 120 patients with castration-resistant metastatic prostate cancer using CellSearch®,[39] and concluded that elevated baseline levels of CTCs was a strong predictor of impaired survival. The prognostic significance of CTCs in prostate cancer was further validated by de Bono and his colleagues, who evaluated CTC counts in the blood from 232 patients with castration-resistant metastatic prostate cancer.[40] Using the cut-off value of $\geq$5 CTCs per 7.5 mL of blood, they showed that CTC counts before and after treatment were significantly associated with survival. Analysis in a subset of patients about to receive first line treatment demonstrated that increased levels of CTCs after treatment was a strong risk factor for death.[41] Thalgott et al. enumerated CTCs in patients at different stages of disease (locally advanced and metastatic prostate cancer) as well as in healthy control individuals and showed that CTC counts in patients with localized disease were similar to that of healthy controls but were significantly much lower compared to those with metastatic disease.[42] Furthermore, they observed significant differences in the levels of CTCs among patients with different metastatic sites.

In addition to epithelial markers, the expression of prostate-specific markers such as prostate-specific antigen (*PSA*), prostate-specific membrane antigen (*PSMA*), and prostate stem cell antigen (*PSCA*) have been used as surrogates for the presence of CTCs in the blood. RT-PCR analysis for the detection of these different tumor markers has been applied in several studies.[43–46] For instance, a study that analyzed both preoperative and intraoperative CTCs using a dual *PSA/PSMA* RT-PCR assay showed that disease-free survival (*DFS*) was shorter in patients with detectable expression of both *PSA* and *PSMA* compared to those who were negative for both markers.[45] In addition, this study also demonstrated that detection of CTCs prior to surgery was a strong prognostic factor for disease recurrence.

5.3.3 Colorectal Cancer

Surgery is currently the definitive treatment for colorectal cancer. After surgery, most colon cancer patients receive adjuvant chemotherapy, while rectal cancer patients undergo chemoradiation therapy. Despite this aggressive treatment, 35–43% of patients eventually experience disease recurrence—mainly liver and lung metastases.[47,48] Recent studies have investigated the use of CTCs to identify patients that have high risk for recurrence. In a study by Uen et al., membrane arrays consisting of a panel of mRNA markers such as human telomerase reverse transcriptase (*hTERT*), cytokeratin-19 (*CK-19*), cytokeratin-20 (*CK-20*), and carcinoembryonic antigen (*CEA*) mRNA were used to detect CTCs in stage I–III colorectal cancer patients who received curative resection for the primary lesion.[49] The presence of CTCs as defined by the expression of all four molecular markers was a strong predictor for post-operative recurrence. A similar study conducted in 735 patients who were candidates for curative surgery showed that the detection of CTCs in patients with advanced stage (Dukes's stage B and C) was associated with high risk of recurrence.[50] These data indicate that CTC detection might help identify non-metastatic colorectal cancer patients that are at high-risk for relapse after surgery.

In the metastatic setting, a prospective multicenter study with 430 patients showed that the detection of CTCs (≥3 CTCs per 7.5 mL by CellSearch®) before and during treatment was an independent predictor of impaired PFS and OS.[51] A similar conclusion was reached in another study involving metastatic patients who received chemotherapy and targeted agents.[52] In addition, a meta-analysis involving over 3000 patients in various stages of colorectal cancer showed that the detection of CTCs indicated poor prognosis.[53]

5.3.4 Lung Cancer

Small cell lung cancer (SCLC) comprises about 13% of all lung cancer cases and is characterized by high mortality rates due to early and rapid spread of the disease with high tropism to the brain. Currently, there are no accurate biomarkers for prognosis or disease monitoring in SCLC. Recent pilot studies have shown elevated CTC numbers (median 50–60 CTCs per 7.5 mL) and a high detection rate (70–90%) in extensive stage SCLC.[54,55] For example, Hiltermann et al. used the CellSearch® system to enumerate CTCs in 59 patients and showed that patients with extensive-stage SCLC had a median of 63 CTCs, while those with limited disease had a median of 6 CTCs per 7.5 mL of blood.[54] The detection of >2 CTCs after first chemotherapy was a strong predictor for poor OS. Naito et al. also used CellSearch® to enumerate CTCs from 51 SCLC patients, and found that 8 CTCs per 7.5 mL was the optimal cut-off value that maximized the hazard ratio (HR) between the unfavorable and favorable group.[56] In a study where CK-19 mRNA-positive CTCs were detected by RT-PCR, the presence of CTCs before and during chemotherapy was a strong predictor of DFS and OS.[57] Furthermore, a meta-analysis involving 440 patients from 7 different studies demonstrated that presence of CTCs as determined by CellSearch® and RT-PCR methods was significantly associated with shorter OS and PFS.[58]

Several groups have analyzed CTCs in patients with non-small cell lung cancer (NSCLC) using different markers.[59–67] A prospective study involving 103 patients who underwent curative lobectomy showed that the detection of CTCs (defined as *CEA* mRNA-positive cells by RT-PCR) prior to surgery was an independent prognostic factor.[67] A study using CK-19 expression to detect CTCs before and after chemo-radiation treatment showed that the presence of CTCs after treatment was a prognostic factor for DFS and OS, but the detection of CTCs before treatment was not.[59] Although CTC presence was significantly associated with shorter OS and PFS in recent meta-analysis involving 1576 patients from 20 studies,[68] the heterogeneity in the methods and experimental designs used in these studies prevented consensus on the clinical utility of CTCs in lung cancer.

5.4 Molecular Characterization of CTC and Personalized Medicine

5.4.1 Tumor Biomarkers in CTCs

With the advent of targeted therapy, evaluation of companion tumor biomarkers has become a prerequisite in choosing optimal treatment modalities. For example, biomarkers in breast cancer such as the estrogen receptor (ER) and the human epidermal growth factor receptor-2 (HER2) have been routinely used to guide hormonal and HER2-targeted therapy, respectively.[69,70] Additionally, the androgen receptor (AR) has shown great potential as a companion tumor biomarker for AR-targeted therapy in castration-resistant prostate cancer patients.[71–74] Testing for *KRAS* status in colorectal cancer plays a key role in clinical decisions involving the application of epidermal growth factor receptor (*EGFR*)-targeted therapy since the presence of KRAS mutation predicted poor response.[75] The use of TKIs (tyrosine kinase inhibitors), such as erlotinib or gefitinib to treat NSCLC, has been guided by information on *EGFR* mutation status.[76] Also, the detection of *EML4-ALK* fusion oncogene has facilitated the selection of patients who are more likely to benefit from ALK kinase inhibitors.[77]

There is a growing body of evidence that biomarker status used to guide targeted therapy can change during disease recurrence or progression.[78–82] Changes in biomarker status, which can play a major role in resistance to treatment have been demonstrated by comparing metastatic lesions with corresponding primary tumors.[79,80] For example, the use of TKIs to target *EGFR* in patients with NSCLC is effective, but most patients experience a relapse within 1 year after the initiation of therapy. On the basis of results from follow-up biopsies, it is thought that acquisition of a secondary *EGFR* mutation causes drug resistance.[81,82] Although up-to-date biomarker status is crucial for targeted therapies, obtaining serial metastatic tissues from distant sites for the purpose of personalized medicine is not feasible.

CTCs, on the other hand, may provide an alternative source of metastatic tissue, which can in turn be used for comparisons with the corresponding primary tumor. Interestingly, recent studies have reported that the HER2 and ER status in CTCs may differ from that of the primary tumor of origin.[83–89] In addition, several studies have tested CTC-related biomarkers in the context of current applications of targeted therapy in the clinic. For example, analysis of important biomarkers, such as ER, EGFR, and HER2 in CTCs has demonstrated heterogeneity among cells from the same patients.[87,90,91] Feasibility studies involving *KRAS* mutation screening in CTCs from early-stage colorectal cancer patients showed high sensitivity and specificity comparable to detection in primary tumor tissues.[92] Moreover, the detection of *EGFR* mutation in CTCs from patients with localized NSCLC showed significant correlation with treatment response.[93]

5.4.2 Molecular Analysis of CTCs

The field of CTC molecular characterization has been very active in recent years.[9] One of the barriers to molecular analysis of CTCs includes the technical difficulties in isolating these extremely rare cells from the overwhelming background of blood cells. The limited amount of nucleic acids available for interrogation presents another obstacle. However, recent advances in rare-cell technology and molecular biology (including reliable nucleic acid amplification techniques) have facilitated the isolation and molecular analysis of CTCs [reviewed in Magbanua and Park[9]]. For example, genome-wide copy number analyses have revealed the extent of genomic instability in CTCs providing evidence of their malignant nature.[94–99] Comparison of genomic profiles between CTCs and their matched primary tumors revealed similarities in chromosomal aberrations indicating clonal relationship.[94,95,98–100] RNA expression analysis has shown the over-expression of well-documented oncogenes such as *ERBB2* in breast CTCs and *AR* in prostate CTCs.[9] Transcriptome analysis, however, has lagged behind DNA-based profiling, partly due to difficulties intrinsic to RNA analysis, such as degradation issues faced when working with limiting amounts of RNA.

5.4.3 Clinical Trials for Personalized Medicine

It has been hypothesized that CTCs may serve as surrogates for the genetics of systemic disease. However, CTC-based biomarkers are currently not clinically actionable. In fact, more studies are needed for CTCs to be fully adopted in the clinic to guide treatment. Recent clinical trials have evaluated the utility of CTCs as a biomarker for response to therapy, some of which have reported promising results.

Although current practice does not consider HER2-negative breast cancers as candidates for HER2-targeted treatment, the application of trastuzumab in patients with HER2-positive CTCs has been hypothesized to confer improvements in clinical outcome. For example, Georgoulias et al. investigated the effect of trastuzumab in HER2-negative breast cancer patients with CTCs that were HER2-positive.[101] Patients who received trastuzumab showed better DFS compared to the observation group suggesting the possible elimination of HER2-positive CTCs by trastuzumab.

A pilot study that analyzed gene expression markers including ERβ[102] revealed that patients with ERβ-positive CTCs experienced a better response to hormonal therapy compared to a subset of patients with ERβ-negative CTCs. This result supports the possibility of using ER expression status in CTCs to help determine which patients are more likely to benefit from hormonal therapy.

Associations between KRAS status in CTCs and treatment response to cetuximab were evaluated in colorectal cancer patients.[103] Results of the clinical trial showed that wild-type KRAS in CTCs was strongly correlated with good response to cetuximab plus chemotherapy. Assessment of mutations in CTCs can therefore be potentially used to select patients for specific targeted therapy.

A study involving chemo-sensitivity profiling in CTCs via expression analysis of a panel of multidrug resistance genes showed high accuracy in predicting response to chemotherapy in a variety of cancers such as bladder, colon, breast, and NSCLC.[104] Chemo-sensitivity was defined as the absence of gene transcription for certain drug transporters related to each type of chemotherapy. For example, undetectable expression of the human equilibrative nucleoside transporter-1 (*hENT*) in CTCs was associated with gemcitabine resistance. In addition, 95% of the patients who experienced disease progression were also predicted to be "resistant to chemotherapy". This novel approach involving chemo-sensitivity assay in CTCs shows great promise for selection of chemotherapy that will provide the most benefit for a particular patient.

5.5 Summary

Molecular characterization of CTCs to guide therapy may become a routine practice in the foreseeable future (Fig. 5.1).[105] CTC profiling may provide a better understanding of the molecular changes associated with tumor progression and metastasis and facilitate the identification of biomarkers for selection of appropriate treatment. In addition, serial analysis of CTCs may facilitate the optimization of treatment decisions based on real-time assessment of tumor characteristics. Finally, validation studies in large clinical trials are warranted in order for CTCs to be accepted in the clinic as biomarkers for personalized medicine.

References

1. Valastyan S, Weinberg RA. Tumor metastasis: molecular insights and evolving paradigms. *Cell.* 2011;147:275–92.
2. Fidler IJ. The pathogenesis of cancer metastasis: the 'seed and soil' hypothesis revisited. *Nat Rev Cancer.* 2003;3:453–8.
3. Ashworth TR. A case of cancer in which cells similar to those in the tumours were seen in the blood after death. *Med J Aust.* 1869;14:146–7.
4. Ross AA, Cooper BW, Lazarus HM, Mackay W, Moss TJ, Ciobanu N, et al. Detection and viability of tumor cells in peripheral blood stem cell collections from breast cancer patients using immunocytochemical and clonogenic assay techniques. *Blood.* 1993;82:2605–10.
5. Naume B, Borgen E, Beiske K, Herstad TK, Ravnas G, Renolen A, et al. Immunomagnetic techniques for the enrichment and detection of isolated breast carcinoma cells in bone marrow and peripheral blood. *J Hematother.* 1997;6:103–14.
6. Racila E, Euhus D, Weiss AJ, Rao C, McConnell J, Terstappen LW, et al. Detection and characterization of carcinoma cells in the blood. *Proc Natl Acad Sci U S A.* 1998;95:4589–94.
7. Ghossein RA, Bhattacharya S, Rosai J. Molecular detection of micrometastases and circulating tumor cells in solid tumors. *Clin Cancer Res.* 1999;5:1950–60.
8. Kalluri R, Weinberg RA. The basics of epithelial-mesenchymal transition. *J Clin Invest.* 2009;119:1420–8.

9. Magbanua MJ, Park JW. Advances in genomic characterization of circulating tumor cells. *Cancer Metastasis Rev.* 2014;33:757–69.
10. Chen Q, Ge F, Cui W, Wang F, Yang Z, Guo Y, et al. Lung cancer circulating tumor cells isolated by the EpCAM-independent enrichment strategy correlate with Cytokeratin 19-derived CYFRA21-1 and pathological staging. *Clin Chim Acta.* 2013;419:57–61.
11. Morgan TM, Lange PH, Vessella RL. Detection and characterization of circulating and disseminated prostate cancer cells. *Front Biosci.* 2007;12:3000–9.
12. Gertler R, Rosenberg R, Fuehrer K, Dahm M, Nekarda H, Siewert JR. Detection of circulating tumor cells in blood using an optimized density gradient centrifugation. *Recent Results Cancer Res.* 2003;162:149–55.
13. Yu N, Zhou J, Cui F, Tang X. Circulating tumor cells in lung cancer: detection methods and clinical applications. *Lung.* 2015;193:157–71.
14. Coumans FA, van Dalum G, Beck M, Terstappen LW. Filter characteristics influencing circulating tumor cell enrichment from whole blood. *PLoS One.* 2013;8:e61770.
15. Nagrath S, Sequist LV, Maheswaran S, Bell DW, Irimia D, Ulkus L, et al. Isolation of rare circulating tumour cells in cancer patients by microchip technology. *Nature.* 2007;450:1235–9.
16. Stott SL, Hsu CH, Tsukrov DI, Yu M, Miyamoto DT, Waltman BA, et al. Isolation of circulating tumor cells using a microvortex-generating herringbone-chip. *Proc Natl Acad Sci U S A.* 2010;107:18392–7.
17. Ozkumur E, Shah AM, Ciciliano JC, Emmink BL, Miyamoto DT, Brachtel E, et al. Inertial focusing for tumor antigen-dependent and -independent sorting of rare circulating tumor cells. *Sci Transl Med.* 2013;5:179ra47.
18. Allard WJ, Matera J, Miller MC, Repollet M, Connelly MC, Rao C, et al. Tumor cells circulate in the peripheral blood of all major carcinomas but not in healthy subjects or patients with nonmalignant diseases. *Clin Cancer Res.* 2004;10:6897–904.
19. Sieuwerts AM, Kraan J, Bolt J, van der Spoel P, Elstrodt F, Schutte M, et al. Anti-epithelial cell adhesion molecule antibodies and the detection of circulating normal-like breast tumor cells. *J Natl Cancer Inst.* 2009;101:61–6.
20. Hosokawa M, Kenmotsu H, Koh Y, Yoshino T, Yoshikawa T, Naito T, et al. Size-based isolation of circulating tumor cells in lung cancer patients using a microcavity array system. *PLoS One.* 2013;8:e67466.

21. Karabacak NM, Spuhler PS, Fachin F, Lim EJ, Pai V, Ozkumur E, et al. Microfluidic, marker-free isolation of circulating tumor cells from blood samples. *Nat Protoc.* 2014;9:694–710.
22. Andreopoulou E, Yang LY, Rangel KM, Reuben JM, Hsu L, Krishnamurthy S, et al. Comparison of assay methods for detection of circulating tumor cells in metastatic breast cancer: AdnaGen AdnaTest BreastCancer Select/Detect versus Veridex CellSearch system. *Int J Cancer.* 2012;130:1590–7.
23. Todenhofer T, Hennenlotter J, Feyerabend S, Aufderklamm S, Mischinger J, Kuhs U, et al. Preliminary experience on the use of the Adnatest(R) system for detection of circulating tumor cells in prostate cancer patients. *Anticancer Res.* 2012;32:3507–13.
24. Cristofanilli M, Hayes DF, Budd GT, Ellis MJ, Stopeck A, Reuben JM, et al. Circulating tumor cells: a novel prognostic factor for newly diagnosed metastatic breast cancer. *J Clin Oncol.* 2005;23:1420–30.
25. Nole F, Munzone E, Zorzino L, Minchella I, Salvatici M, Botteri E, et al. Variation of circulating tumor cell levels during treatment of metastatic breast cancer: prognostic and therapeutic implications. *Ann Oncol.* 2008;19:891–7.
26. Liu MC, Shields PG, Warren RD, Cohen P, Wilkinson M, Ottaviano YL, et al. Circulating tumor cells: a useful predictor of treatment efficacy in metastatic breast cancer. *J Clin Oncol.* 2009;27:5153–9.
27. Nakamura S, Yagata H, Ohno S, Yamaguchi H, Iwata H, Tsunoda N, et al. Multi-center study evaluating circulating tumor cells as a surrogate for response to treatment and overall survival in metastatic breast cancer. *Breast Cancer.* 2010;17:199–204.
28. Consoli F, Grisanti S, Amoroso V, Almici C, Verardi R, Marini M, et al. Circulating tumor cells as predictors of prognosis in metastatic breast cancer: clinical application outside a clinical trial. *Tumori.* 2011;97:737–42.
29. Hartkopf AD, Wagner P, Wallwiener D, Fehm T, Rothmund R. Changing levels of circulating tumor cells in monitoring chemotherapy response in patients with metastatic breast cancer. *Anticancer Res.* 2011;31:979–84.
30. Pierga JY, Hajage D, Bachelot T, Delaloge S, Brain E, Campone M, et al. High independent prognostic and predictive value of circulating tumor cells compared with serum tumor markers in a large prospective trial

in first-line chemotherapy for metastatic breast cancer patients. *Ann Oncol.* 2012;23:618–24.

31. Martin M, Custodio S, de Las Casas ML, Garcia-Saenz JA, de la Torre JC, Bellon-Cano JM, et al. Circulating tumor cells following first chemotherapy cycle: an early and strong predictor of outcome in patients with metastatic breast cancer. *Oncologist.* 2013;18:917–23.
32. Cristofanilli M, Budd GT, Ellis MJ, Stopeck A, Matera J, Miller MC, et al. Circulating tumor cells, disease progression, and survival in metastatic breast cancer. *N Engl J Med.* 2004;351:781–91.
33. Hayes DF, Cristofanilli M, Budd GT, Ellis MJ, Stopeck A, Miller MC, et al. Circulating tumor cells at each follow-up time point during therapy of metastatic breast cancer patients predict progression-free and overall survival. *Clin Cancer Res.* 2006;12:4218–24.
34. Bidard FC, Peeters DJ, Fehm T, Nole F, Gisbert-Criado R, Mavroudis D, et al. Clinical validity of circulating tumour cells in patients with metastatic breast cancer: a pooled analysis of individual patient data. *Lancet Oncol.* 2014;15:406–14.
35. Smerage JB, Barlow WE, Hortobagyi GN, Winer EP, Leyland-Jones B, Srkalovic G, et al. Circulating tumor cells and response to chemotherapy in metastatic breast cancer: SWOG S0500. *J Clin Oncol.* 2014;32(31):3483–9.
36. Lucci A, Hall CS, Lodhi AK, Bhattacharyya A, Anderson AE, Xiao L, et al. Circulating tumour cells in non-metastatic breast cancer: a prospective study. *Lancet Oncol.* 2012;13:688–95.
37. Bidard FC, Belin L, Delaloge S, Lerebours F, Ngo C, Reyal F, et al. Time-dependent prognostic impact of circulating tumor cells detection in non-metastatic breast cancer: 70-month analysis of the REMAGUS02 study. *Int J Breast Cancer.* 2013;2013:130470.
38. Rack B, Schindlbeck C, Juckstock J, Andergassen U, Hepp P, Zwingers T, et al. Circulating tumor cells predict survival in early average-to-high risk breast cancer patients. *J Natl Cancer Inst.* 2014;106.
39. Danila DC, Heller G, Gignac GA, Gonzalez-Espinoza R, Anand A, Tanaka E, et al. Circulating tumor cell number and prognosis in progressive castration-resistant prostate cancer. *Clin Cancer Res.* 2007;13:7053–8.
40. de Bono JS, Scher HI, Montgomery RB, Parker C, Miller MC, Tissing H, et al. Circulating tumor cells predict survival benefit from treatment in metastatic castration-resistant prostate cancer. *Clin Cancer Res.* 2008;14:6302–9.

41. Scher HI, Jia X, de Bono JS, Fleisher M, Pienta KJ, Raghavan D, et al. Circulating tumour cells as prognostic markers in progressive, castration-resistant prostate cancer: a reanalysis of IMMC38 trial data. *Lancet Oncol.* 2009;10:233–9.

42. Thalgott M, Rack B, Maurer T, Souvatzoglou M, Eiber M, Kress V, et al. Detection of circulating tumor cells in different stages of prostate cancer. *J Cancer Res Clin Oncol.* 2013;139:755–63.

43. Bianco FJ, Jr., Powell IJ, Cher ML, Wood DP, Jr. Presence of circulating prostate cancer cells in African American males adversely affects survival. *Urol Oncol.* 2002;7:147–52.

44. Lembessis P, Msaouel P, Halapas A, Sourla A, Panteleakou Z, Pissimissis N, et al. Combined androgen blockade therapy can convert RT-PCR detection of prostate-specific antigen (PSA) and prostate-specific membrane antigen (PSMA) transcripts from positive to negative in the peripheral blood of patients with clinically localized prostate cancer and increase biochemical failure-free survival after curative therapy. *Clin Chem Lab Med.* 2007;45:1488–94.

45. Eschwege P, Moutereau S, Droupy S, Douard R, Gala JL, Benoit G, et al. Prognostic value of prostate circulating cells detection in prostate cancer patients: a prospective study. *Br J Cancer.* 2009;100:608–10.

46. Joung JY, Cho KS, Kim JE, Seo HK, Chung J, Park WS, et al. Prostate stem cell antigen mRNA in peripheral blood as a potential predictor of biochemical recurrence in high-risk prostate cancer. *J Surg Oncol.* 2010;101:145–8.

47. Galandiuk S, Wieand HS, Moertel CG, Cha SS, Fitzgibbons RJ, Jr., Pemberton JH, et al. Patterns of recurrence after curative resection of carcinoma of the colon and rectum. *Surg Gynecol Obstet.* 1992;174:27–32.

48. Sargent D, Sobrero A, Grothey A, O'Connell MJ, Buyse M, Andre T, et al. Evidence for cure by adjuvant therapy in colon cancer: observations based on individual patient data from 20,898 patients on 18 randomized trials. *J Clin Oncol.* 2009;27:872–7.

49. Uen YH, Lu CY, Tsai HL, Yu FJ, Huang MY, Cheng TL, et al. Persistent presence of postoperative circulating tumor cells is a poor prognostic factor for patients with stage I-III colorectal cancer after curative resection. *Ann Surg Oncol.* 2008;15:2120–8.

50. Iinuma H, Watanabe T, Mimori K, Adachi M, Hayashi N, Tamura J, et al. Clinical significance of circulating tumor cells, including cancer stem-like cells, in peripheral blood for recurrence and prognosis in

patients with Dukes' stage B and C colorectal cancer. *J Clin Oncol.* 2011;29:1547–55.

51. Cohen SJ, Punt CJ, Iannotti N, Saidman BH, Sabbath KD, Gabrail NY, et al. Relationship of circulating tumor cells to tumor response, progression-free survival, and overall survival in patients with metastatic colorectal cancer. *J Clin Oncol.* 2008;26:3213–21.
52. Tol J, Koopman M, Miller MC, Tibbe A, Cats A, Creemers GJ, et al. Circulating tumour cells early predict progression-free and overall survival in advanced colorectal cancer patients treated with chemotherapy and targeted agents. *Ann Oncol.* 2010;21:1006–12.
53. Rahbari NN, Aigner M, Thorlund K, Mollberg N, Motschall E, Jensen K, et al. Meta-analysis shows that detection of circulating tumor cells indicates poor prognosis in patients with colorectal cancer. *Gastroenterology.* 2010;138:1714–26.
54. Hiltermann TJ, Pore MM, van den Berg A, Timens W, Boezen HM, Liesker JJ, et al. Circulating tumor cells in small-cell lung cancer: a predictive and prognostic factor. *Ann Oncol.* 2012;23:2937–42.
55. Normanno N, Rossi A, Morabito A, Signoriello S, Bevilacqua S, Di Maio M, et al. Prognostic value of circulating tumor cells' reduction in patients with extensive small-cell lung cancer. *Lung Cancer.* 2014;85:314–9.
56. Naito T, Tanaka F, Ono A, Yoneda K, Takahashi T, Murakami H, et al. Prognostic impact of circulating tumor cells in patients with small cell lung cancer. *J Thorac Oncol.* 2012;7:512–9.
57. Shi WL, Li J, Du YJ, Zhu WF, Wu Y, Hu YM, et al. CK-19 mRNA-positive cells in peripheral blood predict treatment efficacy and survival in small-cell lung cancer patients. *Med Oncol.* 2013;30:755.
58. Zhang J, Wang HT, Li BG. Prognostic significance of circulating tumor cells in small–cell lung cancer patients: a meta-analysis. *Asian Pac J Cancer Prev.* 2014;15:8429–33.
59. Chen TF, Jiang GL, Fu XL, Wang LJ, Qian H, Wu KL, et al. CK19 mRNA expression measured by reverse-transcription polymerase chain reaction (RT-PCR) in the peripheral blood of patients with non-small cell lung cancer treated by chemo-radiation: an independent prognostic factor. *Lung Cancer.* 2007;56:105–14.
60. Hofman V, Bonnetaud C, Ilie MI, Vielh P, Vignaud JM, Flejou JF, et al. Preoperative circulating tumor cell detection using the isolation by size of epithelial tumor cell method for patients with lung cancer is a new prognostic biomarker. *Clin Cancer Res.* 2011;17:827–35.

61. Krebs MG, Sloane R, Priest L, Lancashire L, Hou JM, Greystoke A, et al. Evaluation and prognostic significance of circulating tumor cells in patients with non-small-cell lung cancer. *J Clin Oncol.* 2011;29:1556–63.
62. Sawabata N, Okumura M, Utsumi T, Inoue M, Shiono H, Minami M, et al. Circulating tumor cells in peripheral blood caused by surgical manipulation of non-small-cell lung cancer: pilot study using an immunocytology method. *Gen Thorac Cardiovasc Surg.* 2007;55:189–92.
63. Yoon SO, Kim YT, Jung KC, Jeon YK, Kim BH, Kim CW. TTF-1 mRNA-positive circulating tumor cells in the peripheral blood predict poor prognosis in surgically resected non-small cell lung cancer patients. *Lung Cancer.* 2011;71:209–16.
64. Muinelo-Romay L, Vieito M, Abalo A, Nocelo MA, Baron F, Anido U, et al. Evaluation of circulating tumor cells and related events as prognostic factors and surrogate biomarkers in advanced NSCLC patients receiving first-line systemic treatment. *Cancers.* 2014;6:153–65.
65. Tarumi S, Gotoh M, Kasai Y, Matsuura N, Okuda M, Go T, et al. Innovative method using circulating tumor cells for prediction of the effects of induction therapy on locally advanced non-small cell lung cancer. *J Cardiothorac Surg.* 2013;8:175.
66. Juan O, Vidal J, Gisbert R, Munoz J, Macia S, Gomez-Codina J. Prognostic significance of circulating tumor cells in advanced non-small cell lung cancer patients treated with docetaxel and gemcitabine. *Clin Transl Oncol.* 2014;16:637–43.
67. Yamashita J, Matsuo A, Kurusu Y, Saishoji T, Hayashi N, Ogawa M. Preoperative evidence of circulating tumor cells by means of reverse transcriptase-polymerase chain reaction for carcinoembryonic antigen messenger RNA is an independent predictor of survival in non-small cell lung cancer: a prospective study. *J Thorac Cardiovasc Surg.* 2002;124:299–305.
68. Huang J, Wang K, Xu J, Huang J, Zhang T. Prognostic significance of circulating tumor cells in non-small-cell lung cancer patients: a meta-analysis. *PLoS One.* 2013;8:e78070.
69. Early Breast Cancer Trialists' Collaborative G. Effects of chemotherapy and hormonal therapy for early breast cancer on recurrence and 15-year survival: an overview of the randomised trials. *Lancet.* 2005;365:1687–717.

70. Romond EH, Perez EA, Bryant J, Suman VJ, Geyer CE, Jr., Davidson NE, et al. Trastuzumab plus adjuvant chemotherapy for operable HER2-positive breast cancer. *N Engl J Med.* 2005;353:1673–84.

71. Scher HI, Fizazi K, Saad F, Taplin ME, Sternberg CN, Miller K, et al. Increased survival with enzalutamide in prostate cancer after chemotherapy. *N Engl J Med.* 2012;367:1187–97.

72. Hoffman-Censits J, Kelly WK. Enzalutamide: a novel antiandrogen for patients with castrate-resistant prostate cancer. *Clin Cancer Res.* 2013;19:1335–9.

73. Clegg NJ, Wongvipat J, Joseph JD, Tran C, Ouk S, Dilhas A, et al. ARN-509: a novel antiandrogen for prostate cancer treatment. *Cancer Res.* 2012;72:1494–503.

74. Rathkopf DE, Morris MJ, Fox JJ, Danila DC, Slovin SF, Hager JH, et al. Phase I study of ARN-509, a novel antiandrogen, in the treatment of castration-resistant prostate cancer. *J Clin Oncol.* 2013;31:3525–30.

75. Siddiqui AD, Piperdi B. KRAS mutation in colon cancer: a marker of resistance to EGFR-I therapy. *Ann Surg Oncol.* 2010;17:1168–76.

76. Shepherd FA, Rodrigues Pereira J, Ciuleanu T, Tan EH, Hirsh V, Thongprasert S, et al. Erlotinib in previously treated non-small-cell lung cancer. *N Engl J Med.* 2005;353:123–32.

77. Shaw AT, Kim DW, Nakagawa K, Seto T, Crino L, Ahn MJ, et al. Crizotinib versus chemotherapy in advanced ALK-positive lung cancer. *N Engl J Med.* 2013;368:2385–94.

78. Lindstrom LS, Karlsson E, Wilking UM, Johansson U, Hartman J, Lidbrink EK, et al. Clinically used breast cancer markers such as estrogen receptor, progesterone receptor, and human epidermal growth factor receptor 2 are unstable throughout tumor progression. *J Clin Oncol.* 2012;30:2601–8.

79. Thompson AM, Jordan LB, Quinlan P, Anderson E, Skene A, Dewar JA, et al. Prospective comparison of switches in biomarker status between primary and recurrent breast cancer: the Breast Recurrence In Tissues Study (BRITS). *Breast Cancer Res.* 2010;12:R92.

80. Niikura N, Liu J, Hayashi N, Mittendorf EA, Gong Y, Palla SL, et al. Loss of human epidermal growth factor receptor 2 (HER2) expression in metastatic sites of HER2-overexpressing primary breast tumors. *J Clin Oncol.* 2012;30:593–9.

81. Kobayashi S, Boggon TJ, Dayaram T, Janne PA, Kocher O, Meyerson M, et al. EGFR mutation and resistance of non-small-cell lung cancer to gefitinib. *N Engl J Med.* 2005;352:786–92.

82. Pao W, Miller VA, Politi KA, Riely GJ, Somwar R, Zakowski MF, et al. Acquired resistance of lung adenocarcinomas to gefitinib or erlotinib is associated with a second mutation in the EGFR kinase domain. *PLoS Med.* 2005;2:e73.

83. Punnoose EA, Atwal SK, Spoerke JM, Savage H, Pandita A, Yeh RF, et al. Molecular biomarker analyses using circulating tumor cells. *PLoS One.* 2010;5:e12517.

84. Ignatiadis M, Rothe F, Chaboteaux C, Durbecq V, Rouas G, Criscitiello C, et al. HER2-positive circulating tumor cells in breast cancer. *PLoS One.* 2011;6:e15624.

85. Hayashi N, Nakamura S, Tokuda Y, Shimoda Y, Yagata H, Yoshida A, et al. Prognostic value of HER2-positive circulating tumor cells in patients with metastatic breast cancer. *Int J Clin Oncol.* 2012;17:96–104.

86. Aktas B, Muller V, Tewes M, Zeitz J, Kasimir-Bauer S, Loehberg CR, et al. Comparison of estrogen and progesterone receptor status of circulating tumor cells and the primary tumor in metastatic breast cancer patients. *Gynecol Oncol.* 2011;122:356–60.

87. Babayan A, Hannemann J, Spotter J, Muller V, Pantel K, Joosse SA. Heterogeneity of estrogen receptor expression in circulating tumor cells from metastatic breast cancer patients. *PLoS One.* 2013;8:e75038.

88. Tewes M, Aktas B, Welt A, Mueller S, Hauch S, Kimmig R, et al. Molecular profiling and predictive value of circulating tumor cells in patients with metastatic breast cancer: an option for monitoring response to breast cancer related therapies. *Breast Cancer Res Treat.* 2009;115:581–90.

89. Sieuwerts AM, Mostert B, Bolt-de Vries J, Peeters D, de Jongh FE, Stouthard JM, et al. mRNA and microRNA expression profiles in circulating tumor cells and primary tumors of metastatic breast cancer patients. *Clin Cancer Res.* 2011;17:3600–18.

90. Bock C, Rack B, Kuhn C, Hofmann S, Finkenzeller C, Jager B, et al. Heterogeneity of ERalpha and ErbB2 status in cell lines and circulating tumor cells of metastatic breast cancer patients. *Transl Oncol.* 2012;5:475–85.

91. Gasch C, Bauernhofer T, Pichler M, Langer-Freitag S, Reeh M, Seifert AM, et al. Heterogeneity of epidermal growth factor receptor status and mutations of KRAS/PIK3CA in circulating tumor cells of patients with colorectal cancer. *Clin Chem.* 2013;59:252–60.

92. Yang MJ, Chiu HH, Wang HM, Yen LC, Tsao DA, Hsiao CP, et al. Enhancing detection of circulating tumor cells with activating KRAS oncogene

in patients with colorectal cancer by weighted chemiluminescent membrane array method. *Ann Surg Oncol.* 2010;17:624–33.

93. Maheswaran S, Sequist LV, Nagrath S, Ulkus L, Brannigan B, Collura CV, et al. Detection of mutations in EGFR in circulating lung-cancer cells. *N Engl J Med.* 2008;359:366–77.
94. Heitzer E, Auer M, Gasch C, Pichler M, Ulz P, Hoffmann EM, et al. Complex tumor genomes inferred from single circulating tumor cells by array-CGH and next-generation sequencing. *Cancer Res.* 2013;73:2965–75.
95. Magbanua MJ, Sosa EV, Roy R, Eisenbud LE, Scott JH, Olshen A, et al. Genomic profiling of isolated circulating tumor cells from metastatic breast cancer patients. *Cancer Res.* 2013;73:30–40.
96. Mathiesen RR, Fjelldal R, Liestol K, Due EU, Geigl JB, Riethdorf S, et al. High-resolution analyses of copy number changes in disseminated tumor cells of patients with breast cancer. *Int J Cancer.* 2012;131:E405–15.
97. Hannemann J, Meyer-Staeckling S, Kemming D, Alpers I, Joosse SA, Pospisil H, et al. Quantitative high-resolution genomic analysis of single cancer cells. *PLoS One.* 2011;6:e26362.
98. Magbanua MJ, Sosa EV, Scott JH, Simko J, Collins C, Pinkel D, et al. Isolation and genomic analysis of circulating tumor cells from castration resistant metastatic prostate cancer. *BMC Cancer.* 2012;12:78.
99. Ni X, Zhuo M, Su Z, Duan J, Gao Y, Wang Z, et al. Reproducible copy number variation patterns among single circulating tumor cells of lung cancer patients. *Proc Natl Acad Sci U S A.* 2013;110:21083-8.
100. Ulmer A, Schmidt-Kittler O, Fischer J, Ellwanger U, Rassner G, Riethmuller G, et al. Immunomagnetic enrichment, genomic characterization, and prognostic impact of circulating melanoma cells. *Clin Cancer Res.* 2004;10:531–7.
101. Georgoulias V, Bozionelou V, Agelaki S, Perraki M, Apostolaki S, Kallergi G, et al. Trastuzumab decreases the incidence of clinical relapses in patients with early breast cancer presenting chemotherapy-resistant CK-19mRNA-positive circulating tumor cells: results of a randomized phase II study. *Ann Oncol.* 2012;23:1744–50.
102. Wang H, Molina J, Jiang J, Ferber M, Pruthi S, Jatkoe T, et al. Gene expression markers in circulating tumor cells may predict bone metastasis and response to hormonal treatment in breast cancer. *Mol Clin Oncol.* 2013;1:1031–8.

103. Yen LC, Yeh YS, Chen CW, Wang HM, Tsai HL, Lu CY, et al. Detection of KRAS oncogene in peripheral blood as a predictor of the response to cetuximab plus chemotherapy in patients with metastatic colorectal cancer. *Clin Cancer Res.* 2009;15:4508–13.

104. Gazzaniga P, Naso G, Gradilone A, Cortesi E, Gandini O, Gianni W, et al. Chemosensitivity profile assay of circulating cancer cells: prognostic and predictive value in epithelial tumors. *Int J Cancer.* 2010;126:2437–47.

105. Lee, JS, Magbanua, MJ, Park, JW. Circulating tumor cells in breast cancer: applications in personalized medicine. *Breast Cancer Res Treat.* 2016;160:411–24.

Chapter 6

Mouse Models in Personalized Cancer Medicine

M. E. Beaulieu,[a,*] T. Jauset,[a,*] D. Massó-Vallés,[a,*] L. Soucek,[a,b,c] and J. R. Whitfield[a]

[a] *Vall d'Hebron Institute of Oncology (VHIO), Edifici Mediterrània, 08035 Barcelona, Spain*
[b] *Institució Catalana de Recerca i Estudis Avançats (ICREA), 08010 Barcelona, Spain*
[c] *Department of Biochemistry and Molecular Biology, Universitat Autonoma de Barcelona, 08193 Bellaterra, Spain*
lsoucek@vhio.net

From the identification of the first oncogenes and tumor suppressors to the more recent exploration of high-throughput sequencing data from studies like The Cancer Genome Atlas (TCGA) (Chin et al., 2011; Cancer Genome Atlas Research Network, 2013), mouse models have helped cancer researchers to decipher and disentangle the driving molecular and cellular events that lead to cancer, but also to develop and test increasingly specialized drug-targeting strategies to improve their efficacy and reduce the associated side effects. Indeed, it is now clear that in addition to the genetic aberrations that underlie cell transformation, the microenvironment—formed by surrounding stromal cells and by immune components—plays a critical role in tumor expansion,

*These authors contributed equally.

Cancer Genetics and Genomics for Personalized Medicine
Edited by Il-Jin Kim

ISBN 978-981-4669-87-0 (Hardcover), 978-1-315-36476-6 (eBook)
www.panstanford.com

cancer progression, and response to therapy. To date, despite significant advances in cellular culture techniques (including 3D cultures and co-cultures of tumor and stromal cells), this complexity is not justly recapitulated in vitro.

The implementation of constantly evolving and refined genetic modification technologies in cancer mouse models has enabled dramatic progress in our understanding of the disease. This chapter describes the various types of mouse models of cancer—transgenic, patient-derived xenograft, and humanized models—and provides an overview of their crucial and recent contributions to medical research. From the development of therapies aimed at common alterations in cancer to increasingly targeted treatments based on the particular aberrations found in subsets of patients within a given cancer type, the next foreseen implementation would lead to fully targeted, personalized medicine decisions potentially based on the response of humanized mouse avatars harboring patient-derived tumors to a panel of drugs and combination treatments. This represents a lofty goal. In the shorter term, mouse models continue to help with the progress towards personalized medicine at each step: studying the role of oncogenes and tumor suppressors common to different cancer types and subtypes, preclinical testing of novel targeted therapies, and determining the response of a patient's own tumor to multiple therapies using mouse avatars. Prospective future directions will likely continue to rely on the combination of improved humanized patient-derived xenograft (PDX) models and more elaborate/complex genetically engineered mouse models (GEMMs) to account for the individual character of a patient's tumor.

6.1 Transgenic Mouse Models: From Understanding the Molecular Basis of Tumorigenesis to the Refinement of Drug Design Targeting Specific Mutated/Altered Proteins

GEMMs of cancer consist of mice harboring modifications within their genomes such as mutation and overexpression of oncogenes or inactivation of tumor suppressors that will drive tumorigenesis. By using tissue specific promoters, it is possible to create GEMMs that develop tumors only in the tissue of interest (e.g. pancreas, lung, breast, or skin). An important feature of GEMMs is to faithfully recapitulate the human disease, in terms of how it is initiated, the progressive accumulation of molecular and cellular changes, and remodeling of the tumor microenvironment. Recent advances in the generation of GEMMs of human cancers have consolidated the position of these models as a solid platform for validation of targeted and personalized drugs, offering some advantages that can place these tumors closer to their human counterparts (Singh et al., 2012). These advantages are discussed below along with some examples on how genetic engineering can create models to study cancer development and personalized treatment.

6.1.1 Prenatal GEMMs

GEMMs may be based on the transient activation of tissue-specific promoters during embryonic stages to induce genetic modifications in the tissue of interest. Such alterations will trigger the tumorigenic process from early development and establish tumors at different times during adulthood, depending on the nature of additional, sporadically acquired genetic alterations that are necessary for tumor progression.

6.1.1.1 Tissue-specific promoters and CRE recombinase technology

The first model to faithfully recapitulate the natural progression of pancreatic ductal adenocarcinoma (PDAC) in patients was the *PDX1-Cre;LSL-KRAS*^{G12D} mouse created in 2003 (PDX1: pancreatic and duodenal homeobox 1) (Hingorani et al., 2003). In this model, the constitutively-active KRAS^{G12D} mutant—activating KRAS mutations are found in 90% of human pancreatic tumors (Ryan et al., 2014)—is expressed from the endogenous KRAS locus, as occurs when KRAS is mutated in the human counterpart. Expression of KRAS^{G12D} is, however, silenced by a loxP-STOP-loxP cassette (LSL), which can be removed by CRE recombinase. To drive tissue-specific expression of the KRAS mutant in pancreatic tissue, CRE recombinase is placed under the control of the PDX1 promoter, which is specifically activated in progenitor cells of mouse pancreas. Once the STOP cassette is excised, $KRAS^{G12D}$ mutant protein is expressed and triggers pancreatic tumorigenesis. By 10 months of age, neoplastic ducts predominate over normal ones (Hingorani et al., 2003), forming tumors that recapitulate human pancreatic intraepithelial neoplasias (PanINs), with later appearance of pancreatic ductal adenocarcinomas and invasive lesions. The close homology between human and mouse tumors suggests that this model can be used to accurately study tumorigenesis of the human disease and its response to treatment with specific drugs.

To complement the activation of the $KRAS^{G12D}$ mutation, other genes such as the mutated tumor suppressor $p53^{R172H}$ can also be placed following the LSL cassette. CRE recombinase activity will therefore drive $KRAS^{G12D}$ expression while simultaneously triggering expression of the $p53^{R172H}$ mutant, causing the loss of its tumor suppressive function (Hingorani et al., 2005). The *LSL-KRAS*^{G12D}*/LSL-Trp53*^{R172H}*/PDX1-Cre* model was used to investigate the synergistic effect of nab-paclitaxel and gemcitabine, a combination that recently showed significant improvement over gemcitabine alone in clinical trials (Hoff et al., 2013). The safety profile and limited effect on tumor growth of gemcitabine monotherapy in the mouse were consistent with clinical observations (Frese et al., 2012), reflecting the close similarity of the mouse model to human patients. Furthermore, in this same experimental system, a new mechanism of action of nab-paclitaxel/gemcitabine combination therapy against pancreatic cancer was discovered: it was shown that a reduced cytidine deaminase protein level resulting from nab-paclitaxel treatment increased the gemcitabine concentration in the tumor area, thus improving its antitumor effect (Frese et al., 2012). This notion can be further exploited in the future to design new therapies against KRAS-driven pancreatic cancer.

In fact, GEMMs can also be employed to discover and develop new promising drugs. For instance, the same model was used to demonstrate the inhibition of carcinogenesis displayed by sulindac, an inhibitor of aldo-keto reductase 1B10 (Li et al., 2013) and, in a similar mouse, the promising therapeutic effect of a new anti-Myc drug (Stellas et al., 2014). The close correlation between GEMMs and available human clinical data suggests that these two inhibitors may be effective against KRAS-driven tumors (Singh et al., 2010). Similarly, current mouse models cover many additional alterations found in patients, such as LKB1, Notch and Smad4 (Guerra and Barbacid, 2013), and can be used to precisely anticipate the effects of anticancer drugs in the corresponding human tumors.

6.1.2 Postnatal GEMMS

Spontaneous mouse models such as the one described above are widely used and accepted, but despite the notable communalities of the tumors in patients and GEMMs, they do not entirely recapitulate the human disease. The main difference resides in their distinct etiology: except for hereditary cancer, the majority of cancers are caused by sporadic mutations acting as tumor-initiating events in fully developed (adult) organisms, a phenomenon not accounted for by the "prenatal" GEMMs where these events are produced during early stages of development. Other models where tumors are induced at will in the adult mouse provide more representative alternatives in this regard (Saunders, 2011).

6.1.2.1 Irreversible models: Inducible CRE recombination

A seminal example of this approach published in 2001 described the generation of an inducible non-small cell lung cancer (NSCLC) model (Jackson, 2001). Again, CRE technology is used, although CRE is not present in the genome of the *LSL-KRAS*^{G12D} animals but instead delivered directly to lung cells by intranasal administration of an adenovirus (AdenoCRE). Infection of bronchoalveolar epithelial cells by the virus triggers mutant KRAS^{G12D} expression that elicits the formation of multiple tumors in the lung. This system was used to generate KRAS and BRAF driven lung cancers in which the MEK inhibitor selumetinib (AZD6244) showed great antitumor effect (Ji et al., 2007). A clinical trial confirmed the benefit of selumetinib in combination with docetaxel compared to the docetaxel monotherapy (Janne et al., 2013). Interestingly, while the clinical trials were carried out, results from another study using selumetinib in mouse models were published: the *LSL-KRAS*^{G12D} model of NSCLC was complemented with inactivation of either p53 or LKB1—both tumor suppressor genes that are often mutated in lung cancer—and, while *KRAS*^{G12D} only and *KRAS*^{G12D}*/p53*$^{-/-}$ tumors remained sensitive to selumetinib, *KRAS*^{G12D}*/LBK1*$^{-/-}$ displayed a marked primary resistance (Chen et al., 2012). In the combination treatment clinical trial, mutation in the LKB1 gene was not one of the criteria for recruitment. Therefore, retrospective analysis of LKB1 mutation should be conducted to better refine the results of the trial. Thus, in the future, patient treatment regimens can be modified according to results based on GEMMs: in this case the subset of LKB1 mutant patients should not receive selumetinib. However, the same study showed increased activation of AKT in the *KRAS*^{G12D}*/LKB1*$^{-/-}$ tumors (Chen et al., 2012), opening the door to other rational combinations of docetaxel with PI3K/AKT inhibitors against tumors presenting simultaneous KRAS and LKB1 mutations, further personalizing their treatment.

Numerous other lung cancer mouse models are currently available, including some presenting clinically relevant alterations in PTEN, PI3K, and EGFR (Kwon and Berns, 2013). For instance, models with EGFR mutations reproduce the clinical resistance to EGFR tyrosine kinase inhibitors (Regales et al., 2007). As the GEMMs become more genetically modified to fit patient mutation profiles, more appropriate drug combinations can be predicted and the treatments can become more personalized.

6.1.2.2 Switchable models: The tetracycline inducible system

One of the limitations of CRE-mediated recombination is that it is not reversible. In general, this is not a severe shortcoming in mouse models of cancer because, unfortunately, human cancer is rarely naturally reversible. However, there are cases where oncogenic switchability is desirable, and furthermore, the ability to toggle the expression of an inhibitor is required to model periodic treatment regimes. In this context, the most common genetic tools to model the effect of a potential drug are expression of siRNA, dominant negatives or other competitors. A popular method for achieving switchability, both in vitro and in vivo, is the use of tetracycline-controlled transcriptional activation (Gossen et al., 1995). In this system, addition of tetracycline (or a derivative such as doxycycline) to the drinking water of mice can trigger either expression or repression of the desired DNA sequence integrated into the mouse genome. For this to take place, the second component of the system is also transgenically inserted: rtTA or tTA, known as Tet-On and Tet-Off transactivators. These transactivator genes give versatility to the system because they can be placed under the control of a constitutive promoter such as CMV to drive their ubiquitous expression, or placed downstream of a tissue specific promoter to trigger expression only in certain tissues.

Switchable systems have been used to great effect to demonstrate the potential for inhibition of the Myc oncogene that was for many years considered "undruggable" for several reasons, including the fact that it is necessary for normal tissue proliferation. Myc is a transcription factor that is deregulated in the majority of human cancers and coordinates numerous physiological and tumorigenic processes. In the absence of in vivo Myc inhibitors, it was not possible to investigate the potential therapeutic effect of Myc inhibition, nor the anticipated catastrophic side effects. However, a dominant negative of Myc called Omomyc was used taking advantage of the switchable system and crossed with the *LSL-KRAS* lung tumor model described above (creating a triple transgenic mouse). Using this complex genetic model, it was shown that Myc inhibition eradicated KRAS-driven lung tumors (Soucek et al., 2008; 2013) and that the side effects not only were minimal, but also were completely reversed upon relief of the inhibition (Soucek et al., 2008).

Using this same method, other groups have expressed switchable oncogenes to study cancer resistance, oncogene addiction, and to model the therapeutic impact of compounds that target driving oncogenes. For example, a mechanism of therapy resistance was highlighted using the doxycycline-inducible system to express a p110α catalytic subunit mutant of PIK3CA. Mutations in this domain are found in more than 25% of human breast cancers (Vogt et al., 2007). Tumors were initiated specifically in the mammary gland by using the *MMTV* promoter to drive rtTA expression (Liu et al., 2011). Thanks to this model, mutant PIK3CA^{H1047R} expression can be switched off and the oncogenic pathways that sustain the growth of established tumors investigated. Specifically, repression of the oncogenic stimulus was achieved by removing doxycycline after 4 weeks of continued $PIK3CA^{H1047R}$ expression. As a consequence, all tumors regressed during the first week and 1–2 months later 33% of tumors completely disappeared, indicating a dependency upon continuous $PIK3CA^{H1047R}$ expression. Consistent with clinical data of resistance to PI3K inhibition (Liu et al., 2009), however, 64% of the initial number regrew independently of the PI3K mutant expression and only three out of six analyzed recurrent tumors were sensitive to a pan-Class I PI3K inhibitor, suggesting that some remain addicted to PI3K pathway stimulation while others do not (Liu et al., 2009). Because two of the three insensitive tumors showed overexpression of c-Myc, which functions downstream of the PI3K pathway, the authors suggested this as a potential mechanism of resistance. Indeed, knockdown of c-Myc in recurrent tumors led to lower tumor incidence, suggesting again that Myc inhibition could be a rational strategic approach to fight the tumors of those patients with acquired resistance to PI3K inhibitors (Liu et al., 2009).

6.1.2.3 The estrogen-receptor system

The estrogen-receptor (ER) inducible and reversible system is perhaps the quickest switchable system (Whitfield et al., 2014). It consists of a fusion of the protein of interest and the ER ligand-binding domain sequence. The construct is typically expressed under the control of a constitutive promoter. However, the protein of interest is physically inhibited by the fused ER domain and, although expressed, it remains inactive. Restoration of protein function can be rapidly switched on by administration of tamoxifen, which once metabolized to 4-hydroxytamoxifen binds ER and de-represses the protein of interest (Picard et al., 1988; Littlewood et al., 1995). Once tamoxifen is cleared from the organism, the protein is again deactivated. Activation and deactivation occur faster than in inducible models such as the Tet system, since switchability is not dependent on protein induction and degradation. Indeed, recent studies on tamoxifen kinetics demonstrate that the protein can be activated within 2 hours, and remains active for 24–48 hours after tamoxifen injection (Wilson et al., 2014). Interestingly, the ER switchable system can be merged to the CRE system to create a "switchable irreversible system" in which CRE is activated temporarily upon exposure to tamoxifen to flox out the LSL sequence (Saunders, 2011). For instance, by placing CRE-ER under the control of a tissue specific promoter such as from the *tyrosinase* one, a melanocyte-specific promoter (Bentley et al., 1994), CRE-ER was expressed in the mouse skin. In this model, contact of tamoxifen with the animal skin is sufficient to activate CRE and excise the loxP sequences, thus activating an oncogenic mutant B-RAF protein V600E, the most common genetic alteration in melanoma (Haluska et al., 2006) and deleting the tumor suppressor PTEN, a combination found in 20% of human melanomas (Lin et al., 2008). Consistent with the human disease, the melanomas produced in these mice are able to invade and metastasize (Dankort et al., 2009). Using this model, a synergistic effect of MEK and mTOR inhibitors was shown to reduce tumor growth, a finding later validated in human cell lines and xenografts presenting other alterations in the RAS pathway (Posch et al., 2013). Clinical trials based on these results are currently being conducted.

6.1.3 GEMMs: Benefits and Drawbacks

Cancer GEMMs exhibit various features that mimic patient tumors and have been widely used to study the response to personalized drugs (Singh et al., 2012). As in the human disease, tumor development results from genetic mutations driving the tumorigenic process. This progressive development faithfully recapitulates a tumor environment that is slowly remodeled as the tumor grows. In this regard, for example, extensive tumor fibrosis in pancreatic cancer was shown to act as a barrier for drug access to the tumor, a feature suitably reproduced in GEMMs employed to validate drugs that target the tumor stroma (Massó-Vallés et al., 2015). In addition, the immune system contribution to constant remodeling of the tumor-associated stroma (Gajewski et al., 2013) and to tumor growth and survival is another valuable characteristic of human tumors preserved in GEMMs. Indeed, an increasing number of new therapies are now directed towards components of the immune system (Snook and Waldman, 2013). As GEMMs harbor natural adaptive and innate immune responses, they can reflect adequately the effects of the tested drugs on the immune system and constitute suitable models to test immunotherapies (at least where such a therapy is cross-reactive with mouse tissue). However, this is a double-edged sword, as some therapies fully immune-compatible in humans may not be so in mice due to foreign antigen recognition that may trigger an immune response. For example, in the case of nab-paclitaxel mentioned above, it was shown that a drug formulation based on human albumin produced an immune response in mice, thus limiting the applicability of GEMMs to the study of short-term therapeutic effects (Hoff et al., 2013). An additional limitation of GEMMs resides in the absence of epigenetic alterations, accumulated over years in human tumors, such as DNA methylation (Diede et al., 2013). Furthermore, the low incidence of metastasis in mice (the main cause of cancer-associated deaths in humans) can somewhat limit their applicability in the context of personalized therapy design. Improved GEMMs are continuously being developed to address some of these issues. For instance, a novel and versatile genome editing technology called CRISPR/Cas9 has been recently applied in vivo. While this new system allows the functional characterization of a catalog of oncogenic lesions, it also enables modeling of the wide range of mutational profiles found in patients and their response to specific anti-cancer drugs (Sánchez-Rivera et al., 2014; Platt et al., 2014). Despite some limitations, GEMMs still constitute a reliable and necessary tool in the evaluation of effectiveness and safety of targeted drugs developed and used in the personalized medicine field, producing data that often resembles the clinical outcome.

6.2 Patient-Derived Xenograft (PDX): Tumor Heterogeneity and the Exploration of Avatars as Predictors of Treatment Outcome

The obvious genetic and phenotypic gap between murine and human biological systems could limit the fidelity of GEMMs to recapitulate entirely the human disease. In an attempt to overcome this limitation and in the hope of informing better treatment decisions for individual patients, a different approach has been developed: patient-derived xenograft (PDX) mouse models (also termed patient-derived tumor xenografts or patient-derived tissue xenografts). In this approach, a small portion of a patient's tumor is engrafted directly in an immunocompromised mouse. These PDX models, in addition to enabling the testing of therapies on primary human cancer cells in a more physiological, in vivo environment, hold the promise of facilitating the establishment of a personalized therapeutic regimen that would be designed according to the patient's genomic and histologic profile and preclinically validated in the PDX (also called an "avatar" in this particular context). Using this strategy could unravel possible resistance mechanisms of the patient's tumor and thus indicate novel targeted combinations to overcome them, and hopefully reduce the toxicity and cost associated with current non-targeted and potentially inefficient therapies.

6.2.1 Starting Material

Although the use of PDX models itself is not new (Fiebig et al., 1985; Houghton et al., 1982; Houghton and Taylor, 1978), the recent advances in more powerful and accessible high-throughput gene sequencing technologies have enabled better characterization and promoted their use (Moro et al., 2012). Various methods are currently used to generate PDX collections (reviewed by Tentler et al., 2012; Morton and Houghton, 2007). Typically, a fresh patient tissue sample (in this case a tumor sample) originating from a biopsy or surgical resection is used as the starting material. In addition to these solid materials, malignant ascites fluid and pleural effusions have also been used. After removal of any necrotic tissue, the sample is either fragmented mechanically and directly implanted as pieces, or dissociated chemically into single-cell suspensions prior to implantation. In some cases, Matrigel® (Corning Inc.), human fibroblasts, or mesenchymal stem cells are added to the preparation.

Whereas single-cell suspensions arguably provide for a more homogeneous and unbiased sampling of the tumor, the chemical or enzymatic dissociation process needed to produce them can alter the antigen-staining profile and even sensitize the cells to detachment-induced apoptosis (i.e., anoikis [Zvibel et al., 2002]). Thus, one should minimize dissociation times, in addition to ensuring selection of a live, single-cell population (vs. dead cells or aggregates) by flow cytometry prior to transplant (Williams et al., 2013). In contrast, mechanical section of the tissue sample into tumor pieces (typically of $\sim$3 mm^2) and direct engraftment conceivably preserve better the original tumor architecture and stromal component, although the extent to which the tumor microvasculature can be preserved is debated. The role of the protein and proteoglycan mesh structurally supporting the cells, the extracellular matrix (ECM), in cancer progression is increasingly recognized as being crucial. In fact, the biophysical interplay between tumor cells and ECM has been shown to profoundly influence tumor growth, cell migration, signaling and even drug response (reviewed by Choi et al., 2014). Indeed, increased collagen expression and deposition and overproduction of heparin-sulfate proteoglycans correlate with active tumorigenesis in vivo (Kauppila et al., 1998; Zhu et al., 1995). In addition to this impacting on the differential expression of matrix remodeling enzymes such as MMPs, cathepsins and heparanases, the different orientation observed for collagen I fibers in cancer tissue (more linearized and projecting perpendicularly instead of forming relaxed and non-oriented fibrils) was shown to promote migration of malignant cells (Levental et al., 2009; Provenzano et al., 2006). Other important components of the tumor stroma include cancer-associated fibroblasts (CAFs [Hanahan and Coussens, 2012; Kalluri and Zeisberg, 2006]), which among others contribute to the synthesis of the ECM and generation of oncogenic signals. Thus, understanding the interactions between a tumor and its supporting stroma is likely to become central to the development of new therapies targeting primary and metastatic tumors.

Finally, cancer cell lines (CCLs), which also originally derived from patient tumors, can also be engrafted in recipient mice. They constitute valuable tools (Abaan et al., 2013), enabling monitoring of tumor growth through more direct imaging and allowing the modeling of novel potential targeted therapies through genetic engineering of the tumor cells themselves. Indeed, by allowing clonal expansion and manipulation in vitro (infection with lentiviruses, fluorescent markers, luciferase, etc.), they enable the application of one of the 3 "Rs" of humane animal experimentation: replacement. Much can be learned from these cells in culture prior to in vivo work. However, these models lack fidelity to the original tumor. Indeed, CCLs often derive from metastasis and might not be informative of the primary tumors they originate from. Also, they result from the selective pressure imparted by in vitro culture conditions and manipulations (e.g., oxidative stress and enzymatic treatments [Williams et al., 2013]). As a result, genotypic and phenotypic profiles are known to be profoundly altered (Daniel et al., 2009), resulting for instance in an altered invasive capacity and artificial reliance on specific growth and survival pathways, possibly lowering their predictive value. In fact, genome-wide gene-expression analysis demonstrated that PDXs preserve global pathway activity and key genetic profile of the patient's primary tumor to a better extent than a corresponding cell line derived from that same patient, even after several mouse generations (Hollingshead et al., 2014; Reyal et al., 2012; Daniel et al., 2009).

6.2.2 Mode of Implantation

The implantation of tumor fragments or cells is either performed orthotopically relative to the patient's primary tumor, or heterotopically, generally subcutaneously or under the kidney capsules. Subcutaneous xenograft represents the easiest type of implantation, and the most widely used. However, it poorly reflects the natural tumor environment and rarely produces metastasis. Kidney capsule engraftment is technically easier than orthotopic, and since it provides a more highly vascularized environment, it constitutes a more receptive implantation site than subcutaneous. In consequence, shorter time to engraftment is often observed relative to subcutaneous implantation, an important factor for studies seeking to implement real-time PDX analysis in personalized cancer treatment. Orthotopic tumor implantation allows the tumor to grow in a more natural anatomic microenvironment with regards to the pH, oxygen, nutrient, and hormone conditions, and therefore often mimics more accurately the original human tumor when analyzing histology, gene expression profiles and metastatic potential compared to heterotopic transplants (Rubio-Viqueira and Hidalgo, 2009). For instance, orthotopic implantation was shown to drastically accelerate the appearance of metastasis in a model of PDAC (Loukopoulos et al., 2004). While this approach substantially improves tumor take rates, in some cases it is simply a requirement, for instance for growth of metastatic nonseminomatous testicular germ cell tumors (Douglas et al., 2001). However, orthotopic xenografts generally require more elaborate surgical protocols and technical skills to be achieved than heterotopic xenografts.

6.2.3 Recipient Mouse Strains

Several strains of immunocompromised mice can be used for implantation, the most commonly used mice being athymic/nude (deficient in T cells), severe combined immune deficient (SCID, both B and T cell deficient), non-obese diabetic (NOD)-SCID, and NOD-SCID gamma (NSG)—the most highly immunodeficient mouse currently available, lacking B cells, T cells, and NK cells in addition to being deficient in dendritic cells and macrophages. The use of these xenograft hosts for various tumor types is reviewed by Hidalgo et al. (2014). Typical take times range from 1 to 4 months, although growth of the grafted tumor is highly dependent on the location of implantation and of tumor type. For instance, successful engraftment rates are generally high for colorectal (75%), ovarian (65%), melanoma (59%), and lung (50%) tumors, while historically much lower for breast tumors (27%, all subtypes included).

To date, no systematic study comparing take rates, genetics, histology or chemosensitivity of the recipient mouse strains has been performed (Hidalgo et al., 2014), although the recent development of NSG clearly raised the statistics of take rates, particularly in tumors difficult to engraft. Note however that NSG mice do not always increase engraftment rates: ER-positive human breast tumors showed little difference in NOD-SCID versus NSG mice, whereas host hormonal supplementation with estradiol resulted in a tenfold increase in take rate (Hidalgo et al., 2014). Interestingly, tumors that were ER-negative, PR-negative and HER2-

negative (triple negative) grew at the fastest rate, consistent with the aggressiveness of the disease. Similarly, metastatic melanoma cells also displayed higher engraftment rates than primary tumor samples (Sivanand et al., 2012; Némati et al., 2010), pointing to engraftment rate as a potential prognostic factor. In a study to determine the potential effect of the host strain on response to therapy, human CCLs (gastric, NSCLC, myeloblastoma, T-cell lymphoma and breast) were engrafted into athymic versus SCID or Rag2−/− mice, and a significant difference in the response to various chemotherapies was observed (Yoshimura et al., 1997), underlying the possible influence of host genetic background on the therapeutic response.

Once it reaches an appreciable size, the tumor growing in this first generation of mice harboring the patient-derived material (F0) can be resected, implanted, and expanded in subsequent generations (F1, F2, F3, etc.). It is often the third generation (F3) that will serve for drug treatment studies. Similarly to GEMMs, several tumor-specific PDX models of colorectal, breast, pancreatic, ovarian, lung, and brain derived tumors have now been established (reviewed by Hidalgo et al., 2014; Tentler et al., 2012). Studies comparing PDX models of NSCLC, SCLC, PDAC and breast cancer among others with their respective donor tumor demonstrated that PDX tumors can maintain the global molecular and genetic signature (Jin et al., 2010) as well as histological heterogeneity, metastatic potential and site of metastasis (DeRose et al., 2011; Daniel et al., 2009; Fichtner et al., 2008; Rubio-Viqueira et al., 2006) and drug responsiveness, at least in the early passages (Hidalgo et al., 2014).

6.2.4 PDXs in Personalized Medicine

Undeniably, analysis of the fidelity of PDX pathology and response compared with the clinical outcome of the donor patients is of the utmost importance. When this can be performed in parallel, the mice are termed "avatars" and their use paves a way towards personalized medicine for patients with cancer, in which novel drug combinations and biomarker identification can be assessed, and potential resistance mechanisms uncovered (summarized in

Fig. 6.1). The following section describes selected examples of the use of such "avatar mice" in personalized medicine in oncology.

To a large extent, studies on PDAC, NSCLC, breast and colorectal cancer PDXs (reviewed by Hidalgo et al., 2014) evidenced response rates comparable with clinical responses to commonly used cytotoxic (i.e., paclitaxel, gemcitabine, 5-FU or irinotecan) or targeted agents. For instance in colorectal cancer (CRC), where there is a long history of success in the establishment, maintenance and study of PDX models (Giovanella et al., 1989), analysis of tumor response to the EGFR inhibitor cetuximab in unselected PDX models showed a 10.6% response rate (Bertotti et al., 2011), concordant with the statistic observed in the clinic. Interestingly, genetic characterization of nonresponsive PDX mice uncovered compensatory pathways through HER2 amplification in a specific subset, and the hypothesis of a possible therapeutic advantage of cetuximab and pertuzumab (anti-HER2 antibody) combination was assessed in a corresponding cohort of xenopatients. Although a recent clinical study demonstrated the therapeutic potential of the combination in humans, the overlapping toxicities unfortunately prejudice further exploration of this therapeutic avenue (Rubinson et al., 2014).

In an orthotopic PDX model of glioblastoma multiforme (GBM), biopsy samples are sectioned and cultured in vitro in agar to form spheroids, which are then implanted in athymic mice with 96% success rate of engraftment (Wang et al., 2009). The resulting PDX model displays several phenotypical characteristics of human GBM, namely vascularization, glioma cell invasion, and necrosis, while conserving the patient's genetic signature. While treatment of this PDX model with anti-VEGF therapy (bevacizumab) indeed reduced tumor volume and vascularity, it unfortunately produced a more invasive phenotype seemingly dependent on PI3K/Akt signaling. Thus, a combinations with targeted compounds to the PI3K pathway (Kurtz and Ray-Coquard, 2012) or against downstream conduits (Annibali et al., 2014) could provide a more effective therapy.

In PDAC, where an intense desmoplastic reaction is characteristic, the use of CCL xenografts generally growing as homogeneous masses of tumor cells with minimal stromal infiltration does not faithfully recapitulate the disease, and orthotopic PDX PDAC

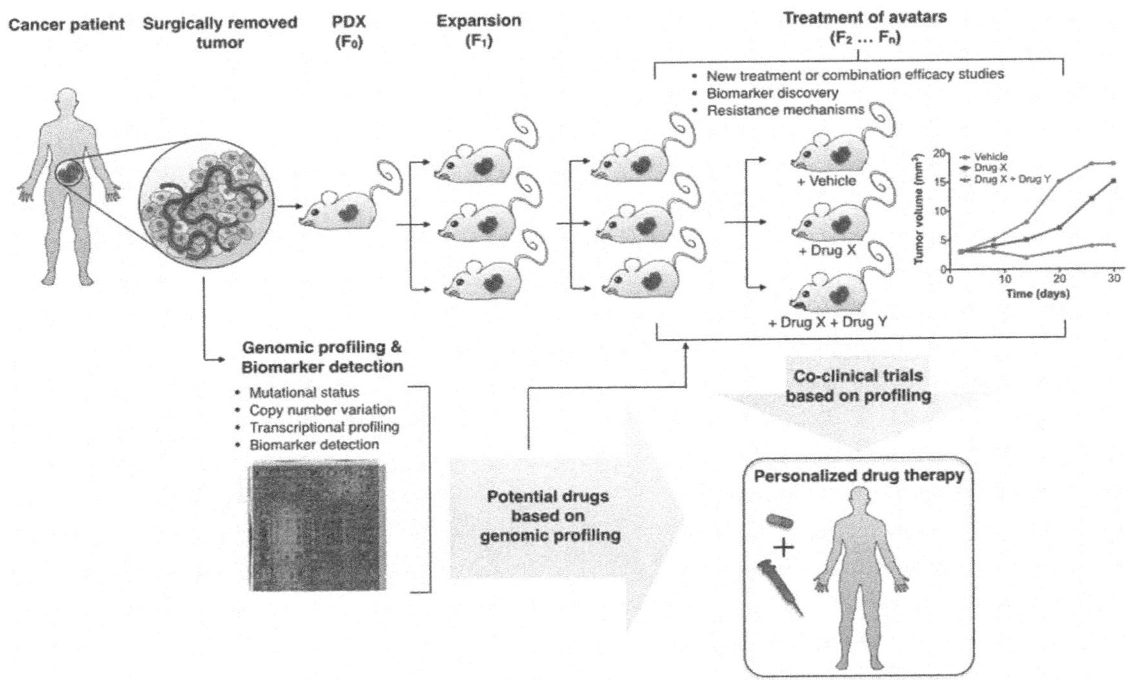

Figure 6.1 Adapted from Tentler et al. (2012) and Hidalgo et al. (2014).

models provide a clear predictive advantage. The use of PDX models combined with whole-genome sequencing, as an empirical screening platform while the patient is receiving first-line therapy, could provide more efficient therapeutic options at the time of disease progression. This strategy demonstrated for instance the activity of mitomycin C and cisplatin in a patient harboring a PALB2 mutation (Jones et al., 2009). As another striking successful example of the use of PDX models as a personalized medicine screening platform, five mouse avatars were generated for each one of 14 refractory pancreatic cancer patients and treated with a total of 63 different drugs in 232 treatment regimens. Analysis of the PDX response enabled identification of 17 prospective guided treatments, of which 15 conferred durable partial remission (Hidalgo et al., 2011). For example, a patient who had been initially treated with epirubicin-cisplatin-capecitabine showed a partial response that lasted 8 months. Subsequent disease progression resulted in lung and liver metastasis, at which point the mouse avatar had responded to the combination irinotecan-bevacizumab-cetuximab. The patient treated with this regimen achieved a partial response of 14 months for the liver metastasis. As the patient's carcinoembryonic antigen (CEA) tumor marker level later started to increase, data from the xenopatient indicated susceptibility to nab-paclitaxel in combination with several angiogenesis inhibitors, which in the patient resulted in normalized CEA levels that were maintained for 8 months (Hidalgo et al., 2011).

It is foreseeable that multi-institution studies will eventually allow for several more extensive genomic, proteomic and metabolomic characterizations of the patient tumor at different time points and successful application of the PDX models to drug discovery and personalized medicine. The Breast Cancer Genome-Guided Therapy (BEAUTY) study, currently carried out by the Mayo Clinic with samples from 200 early-stage, high-risk breast cancer patients, constitutes such an attempt to tailor novel therapies to the patient profile (Boughey et al., 2014). However, a nonscientific limitation of the extensive use of PDXs in current clinical studies is the cost associated with their development and maintenance, with an average cost of $10,000 per patient currently not covered by insurance (Malaney et al., 2014). Harmonization and standardization of study design and data analysis along with information sharing across the various centers developing and routinely using PDX models (with proper confidentiality and data protection systems) would enable faster assessment of model availability, execution of multicenter preclinical trials under a single protocol, and rapid data generation. One such initiative currently being put in place in Europe is the EuroPDX, which aims to create a network of clinically relevant and annotated models of human cancer, in particular PDX models, with the overarching goal to constitute a collection reproducing the heterogeneity of human cancer (Hidalgo et al., 2014).

6.2.5 PDXs: Limitations and Value

An important scientific limitation of PDX models is the inherent absence of a competent immune system. As a result, the actual intricacies of different cell types and cytokines found in the original human tumor is poorly reflected and precludes the use of PDXs to assess immune system targeting therapies. Indeed, active participation of the immune system in creating a favorable niche for cancer maintenance and progression is underscored by the numerous novel therapeutic strategies targeting immune components (e.g., anti-PD1, anti-CD40 antibodies). In addition, the secondary immunomodulatory properties of chemotherapeutics clearly cannot be accounted for. Another intrinsic limitation of xenograft models is the fact that the tumor samples used in their generation often derive from a small portion of the tumor, possibly not representative of the whole mass. When implanted in the mouse, such samples face different selection pressures so that those cells capable of surviving in the new environment and growing most rapidly will "win" the competition and predominate in the newly formed tumor. This bias does not necessarily reflect the development of the tumor in its original context. Hence, lack of heterogeneity in the engrafted tumor is a caveat in the use of PDXs, although it should be noted that many papers have described faithful recapitulation of molecular and pathological profiles between the derived xenograft and the patient tumor (Reyal et al., 2012; DeRose et al., 2011; Moro et al., 2012; Tentler et al., 2012). Finally, more detailed gene expression profiles and histologic analysis showed that the human stroma is gradually replaced with murine stroma. While growth of the human-derived xenograft can be well supported by murine stromal components, this invariably results in key differences in the signaling ligand repertoire potentially critical to the tumor phenotype. Despite these limitations, PDX models have to date proven tremendously useful and informative in the elucidation of fundamental mechanisms of human tumor biology and resistance to therapy, and undeniably constitute a valuable tool in the quest for improved personalized treatments.

6.3 Humanized Mouse Models: Bringing the Major Players Back into the Biological Game

The relationship between immunity and cancer dates back to 1909, when Paul Ehrlich hypothesized that the incidence of cancer would be much higher without the control mechanisms exerted by the immune system (Kaufmann, 2008). When Hanahan and Weinberg revised the Hallmarks of Cancer 100 years later, they included the avoidance of immune destruction—mainly by B and T lymphocytes, macrophages, and natural killer (NK) cells—as an emerging hallmark. They also stressed the tumor-promoting consequences of inflammation, which help incipient tumors to acquire hallmark capabilities (Hanahan and Weinberg, 2011). Immune cells are indeed crucial players in many aspects of tumorigenesis, including cancer initiation, progression, colonization of secondary organs, and responses to therapy (Inman et al., 2014; DeNardo et al., 2008; Qian and Pollard, 2010; Yang et al., 2013).

6.3.1 The Need for Better Models

Patient-derived xenografts (PDXs) have proven to be valuable tools for the study of personalized medicine in cancer and have led to crucial advances in preclinical drug validation. They retain both the intertumoral and intratumoral heterogeneity of the original tumors, as well as their histological architecture and mutation profile (DeRose et al., 2011). However, a major limitation is the requirement to use immunocompromised mice for successful grafting and subsequent growth of human tumors. Given the crucial role that the immune system plays in cancer, there is a need to pursue alternate and more sophisticated models that not only recapitulate patient tumor biology but also predict more accurately the patient's response to therapy and allow validation of immune-based approaches. Immunotherapy has been a major focus of study in recent decades and consists of boosting the immune system to improve, target, or restore its function against cancer cells by several means. These strategies can be classified into three main categories: transfer of immune effectors generated in vitro, such as monoclonal antibodies (mAb) and chimeric antigen receptor T cells; vaccination with whole tumor cells, protein, peptide, or dendritic cells (DC); and immunomodulatory therapy (enhancing T-cell activation, checkpoint blockade and in situ immunization) (Makkouk and Weiner, 2015).

6.3.2 "Humanizing" the Mouse Models

A humanized mouse is a mouse that carries human genes, cells or tissues. The focus of this section will be humanized immune system models, which were initially designed for studying human hematopoiesis and immunity, but are now a promising translational model for the validation of personalized approaches for cancer treatment. Humanized immune system mouse models can be obtained by two main strategies: insertion of human genes into the mouse genome (i.e. human leukocyte antigen [HLA] and cytokines) or reconstitution of the human immune system by engraftment with human hematopoietic and lymphoid cells or tissues, or a combination of both.

The use of highly immunodeficient mice is key for the successful transplantation of both tumor and immune cells, and the introduction of mutated IL-2 receptor gamma chain (*IL-2Rγ*) into established immunodeficient models (NOD/SCID and BALB/c RAG1/$2^{-/-}$) has constituted an important step forward, and indeed critical for immune cell transplantation. IL-2Rγ is a component of the high-affinity receptors for IL-2, IL-4, IL-7, IL-9, IL-15 and IL-21 and is required for their signaling (Sugamura et al., 1996). Three main types of severely immunocompromised mice suited for human immune cell transplantation have been generated: NOG (NOD/Shi-*scid IL-2Rγ*^{null}), NSG (NOD/LtSz-*scid IL-2Rγ*$^{-/-}$) and BRG (BALB/c *Rag1/2*$^{-/-}$ *IL-2Rγ*$^{-/-}$) (reviewed by Ito et al., 2012). The resulting animals lack B, T, and NK cells and have reduced functions of macrophage and DC (Ito et al., 2002), features that allow better engraftment, faster growth, and increased metastatic capacity of solid tumor xenografts compared to the parental IL-2Rγ positive strains (Agliano et al., 2008; Milsom et al., 2013).

6.3.3 Transplantation of Hematopoietic Stem Cells

The IL-2Rγ-deficient background of NOG, NSG, and BRG mice allows proper differentiation of multilineage hematopoietic cell populations after pre-irradiation with sublethal γ-radiation and transplantation of CD34+ human hematopoietic stem cells (HSCs) (Yahata et al., 2002; Hiramatsu et al., 2003). IL-2Rγ$^{-/-}$ mice transplanted with HSCs are able to produce most types of hematopoietic cells, including B and T lymphocytes, NK cells, monocytes, DC, erythrocytes, and platelets (Traggiai et al., 2004; Ishikawa et al., 2005). The use of newborns is preferable, since their engraftment rate is higher when compared to adult mice. T cells colonize the spleen 3–5 months after HSC inoculation (Hiramatsu et al., 2003), while B cells can be detected in peripheral blood 1 month after transplantation and their number increases over time during the first 3 months (Ito et al., 2012). HSCs can be obtained from umbilical cord blood, bone marrow, G-CSF-mobilized peripheral blood and fetal liver and require the appropriate Institutional Review Board approval (Pearson et al., 2008).

NSG and NOG mice have increased HSC engraftment capacity compared to the BRG background, mainly in human T cells (Brehm et al., 2010). This is caused, at least in part, by a polymorphism in the gene encoding the signal regulatory protein alpha (SIRPα) receptor in the NOD background, which allows enhanced binding to the human CD47 ligand, whose expression on mouse macrophages supports human hematopoiesis (Takenaka et al., 2007). Those mice can be further "humanized" by engraftment with fetal thymus and liver from the same source in the kidney capsule. This model is known as NOD/SCID-hu BLT mice (abbreviated BLT) and is currently the humanized mouse model with the most functional immune system. Unlike mice transplanted with HSCs only, BLT mice present human mucosal immunity (Melkus et al., 2006).

6.3.4 Transplantation of Peripheral Blood Mononuclear Cells

Transplantation of human peripheral blood mononuclear cells (PBMCs) into NOG, NSG, and BRG mice is an alternative method for humanization. This is an easier approach than the previous ones, since PBMCs can be isolated from peripheral whole blood of volunteers that sign an informed consent, allowing for higher availability than HSCs. Also, transferred lymphocytes are already functionally mature and studies can be performed in significantly less time compared to HSC-transplanted models (Pearson et al., 2008). Still, their main limitation is the short lifespan of PBMCs in the mouse and the appearance of graft-versus-host disease (GVHD) only 2–4 weeks after injection (van Rijn et al., 2003). Hence, PBMCs are a good source of immune lymphocytes for the study of GVDH but still offer a limited window of opportunity for cancer studies.

6.3.5 Local "Humanization"

Other approaches have been developed to allow local humanization. Subcutaneous implantation of normal breast tissue and inoculation of breast cancer cells on top of these implants adequately mimics the human mammary microenvironment (Wang et al., 2010) and implantation of tissue-engineered human bone grafts allows selective metastasis to the bone (Thibaudeau et al., 2014; Hesami et al., 2014). Thanks to these new humanized models, it is now possible to study bone metastasis in vivo, a common feature of breast and prostate cancer which had not been recapitulated by conventional PDX or GEMMs (Holzapfel et al., 2013).

6.3.6 Overcoming Limitations of the Model

Humanized mice still face several limitations. Despite the presence of most immune cell types after "humanization," the differentiation stage and activity of some of them is sub-optimal, possibly because mouse cytokines are unable to act on human receptors, depriving the engrafted human cells of critical survival and growth factors (Willinger et al., 2011). Additionally, since no HLA molecules are expressed on mouse thymic stromal cells, human T cells that mature in the mouse are unable to recognize antigens in an HLA-restricted manner. Furthermore, IL-2Ry deficient mouse strains lack Peyer's patches and lymph nodes, structures that are necessary for support of the human immune system (Bassuk and Leiden, 1995). Some of these pitfalls are already being addressed and humanized mouse models have been further improved by cytokine knock-in gene replacement (reviewed by Willinger et al., 2011), and addition of a HLA class I heavy and light chain transgenes or a chimeric HLA class II transgene to NSG mice, which resulted in HLA-restricted human immune responses (Shultz et al., 2010; Danner et al., 2011). A recent humanized mouse model derived from BRG mice has overcome some of its current limitations: by knocking in four human cytokines (M-CSF, interleukin 3, thrombopoietin and GM-CSF), MITRG and MISTRG mice were generated (the latter carrying also a SIRPα transgene to improve human cell engraftment). The resulting mice transplanted with HSCs present a higher engraftment efficiency compared to the BRG strain, without requiring irradiation, and with more differentiated populations of hematopoietic cells (Rongvaux et al., 2014).

6.3.7 Success Stories

The highly translational potential of humanized mice has already been extensively demonstrated for many diseases, including cancer. The main use of humanized models has been the study of infectious diseases (Brehm et al., 2013), especially HIV (Denton and Garcia, 2011), and several inflammatory disorders. NSG mice engrafted with human PBMCs derived from patients suffering from atopic dermatitis or healthy volunteers reflected the immunological history of the donors, proving that the model has the potential to provide data that is highly translatable to patients (Zadeh-Khorasani et al., 2013). Regarding cancer, many examples have demonstrated the capacity of humanized mice to mimic key aspects of the disease. NSG mice transplanted with pediatric B-cell precursor acute lymphatic leukemia (B-ALL), myeloid leukemia (AML) and T-cell leukemia (T-ALL) retain their characteristic aberrant immune-phenotype and correlate to the particular clinical outcome of the patients, showing that they are a valuable tool to design and test personalized treatment options for pediatric leukemia patients (Woiterski et al., 2013). MISTRG mice transplanted with HSCs and the melanoma cell line Me290 generated tumors with myeloid cell infiltration, especially from macrophages, which is a common feature of human tumors (Rongvaux et al., 2014). Besides this melanoma model, other humanized mouse models have been generated by co-transplantation of cancer cell lines and HSCs or PBMCs, including Hodgkin lymphoma (HL), cutaneous T-cell lymphoma (CTCL) (Ito et al., 2009), rhabdomyosarcoma (Seitz et al., 2010), breast cancer (Wege et al., 2011), and ovarian cancer (De Giovanni et al., 2012).

In summary, there are a growing number of humanized mouse models that resemble more and more the behavior of human cancer and the crosstalk between the tumor and the immune system, although the use of these models for preclinical validation of immunotherapeutic approaches against cancer has not yet been generalized. Nevertheless, there are already a few examples that showcase the potential of humanized mice in the advancement of targeted therapies that can lead to personalized treatment for cancer. For instance, a chimeric defucosylated anti-CCR4 mAb displays potent antitumor activity in humanized mouse models of HL and CTCL that are transplanted with PBMCs from healthy donors. Tumor regression was mediated by robust antibody-dependent cellular cytotoxicity (ADCC), while no effect was observed in the absence of human immune cells (Ito et al., 2009). Importantly, the human version of the antibody (mogamulizumab) has already shown promising results in clinical trials (Remer et al., 2014). In addition, BRG mice engrafted with HSCs and vaccinated with human HER2+ SK-OV-3 ovarian cancer cells showed human immune anti-HER2 responses and reduced lung metastases after a subcutaneous SK-OV-3 cell challenge, when compared to nonvaccinated animals (De Giovanni et al., 2012). Finally, NSG mice engrafted with human PBMCs allowed in vivo validation of a new technique for DC-based immunotherapy in AML, consisting of antigen loading of DCs via electroporation with in vitro-transcribed mRNA encoding leukemia-associated antigens. This method showed efficient induction of antigen-specific T-cell immune responses and enabled the initiation of a clinical phase I/II trial to test its application in AML patients (reviewed by Subklewe et al., 2014).

6.3.8 Humanized Mouse Models: Future Perspectives

The future of humanized models is now moving towards co-transplantation of a patient's tumor and their own PBMCs or HSCs, and new studies are already under way in this regard. A clear example is a recent study in which NSG mice were humanized by injection of hematopoietic stem and progenitor cells from bone marrow aspirates extracted from carcinoma patients, a procedure that is already performed to assess cancer spread at the time of primary surgery (Werner-Klein et al., 2014). This kind of advance opens the door to the generation of avatar "double-humanized" mouse models that promise to be a powerful tool for validation of immunotherapy-based drugs. More importantly, these double-humanized mice could be used for each patient to develop a cohort of avatars for testing several therapeutic approaches in real time and, in parallel, to determine a personalized primary treatment plan and even anticipate the appearance of resistance and select secondary treatments after tumor relapse or metastasis.

6.4 Discussion and Perspectives

Although investment in R&D of new drugs has increased, there has been no consequent increase in approval rate (Hutchinson and Kirk, 2011; Scannell et al., 2012). The time and cost required to develop an approved new drug has increased dramatically, and cancer drugs are no exception. One reason for this is that despite the great number of drugs entering clinical trials, few are considered safe and effective enough to be used in clinical practice (Hutchinson and Kirk, 2011). Therefore, the preclinical testing systems of candidates prior to patient trials must be optimized to more accurately predict clinical response to drugs and to shoot down more candidates before entering phase I (Denayer et al., 2014; Henderson et al., 2013; Hooijmans et al., 2011).

Mouse models of cancer are the gold standard tool to preclinically assess the safety and efficacy of new cancer drugs and an understanding of what each type of mouse model offers (summarized in Table 6.1) is critical to fully evaluate the most promising drugs that will succeed in clinical trials (Herter-Sprie et al., 2013). GEMMs have been developed that harbor the specific mutations, overexpression of oncogenes or deletion of tumor suppressors found in particular cancers. Such GEMMs increase our understanding of tumorigenesis and aid in the search for targeted therapies. A complementary approach to personalization is to test directly the response of a patient-derived tumor sample to a medication cocktail tailored to its specific mutational profile and use this information to provide a personalized treatment regimen (a "bedside to bench to bedside" approach). Studies have already employed mouse avatars to test panels of drug treatments against the patient's own tumor in order to provide feedback to the clinic to direct the actual course of treatment. This perhaps represents the latest embodiment of a personalized medicine approach. A current focus is on humanization of the immune system, although perhaps in the future other aspects of tumor biology—growth factors, hormones, neovasculature, genomic instability, alterations of metabolic pathways (Hanahan and Weinberg, 2011), or even microbiology (Iida et al., 2013)—could also require humanization to develop a more accurate avatar (Doroshow and Kummar, 2014). In addition, development of models based on circulating tumor cells instead of surgical resection or biopsies constitutes another foreseeable improvement that would enable monitoring of tumor evolution of an individual patient's cancer.

One current limitation common to all mouse models is that some therapeutic regimens used in the clinic simply cannot be recapitulated in mice: for instance, carefully controlled, continuous

Table 6.1 Adapted from Denayer et al. (2014)

Model	Advantages	Disadvantages
Genetically engineered mouse models	• Enable modeling of tumor development and premalignant neoplastic stages • Tumor heterogeneity (multiple independent lesions in one mouse) • Tumor microenvironment (stroma and immune cells) is representative of the studied tumor • Studies on defined mutations possible, including analysis of the effects of these mutations in many genetic backgrounds • Allow reversible expression of the biological (target) oncogene or modeled inhibitor(s) and its tissue-specific expression • Good predictive value regarding therapeutic response and development of resistance	• Metastatic pattern can differ between mouse and human • Lack epigenetic alterations found in human tumors • Tumor and microenvironment are murine and so human/mouse cross-reactive compounds may be required • Development is costly and time consuming • Limited availability of tissue specific promoters
Cell line derived xenografts	• Allows a rapid analysis of response to a therapeutic regimen • Source of material virtually unlimited using immortal cell lines	• In vitro culturing induces artificial drift and leads to outgrowth of cells with different characteristics than primary tumors • Loss of tumor heterogeneity • Human tumor stroma is not represented • Immunodeficient mice cannot adequately capture the intact human immune component • Poor predictive value

Patient-derived xenografts	• Minimal in vitro culture preserves better the molecular and genetic signature of the original tumor and its heterogeneity • Stromal component is representative of the parental tumor in the initial passages • Each patient can have multiple avatars • Offers the opportunity to evaluate tumors from metastatic sites or tumors that have developed resistance to multiple treatments • Studies have shown very good correlation between response in PDX models and clinical response in patients	• Immunocompromised mice cannot adequately capture the intact human immune component • Tumor development in animals is often slow (tumor graft latency between 1 and 12 months) • Low engraftment rates for some tumor types and only a limited source of original material is available • Orthotopic implant is often technically complicated • Expensive and labor intensive to establish and maintain PDX bank
Humanized immune system models[1]	In addition to the advantages of PDXs: • Presence of a human immune system better reflects human tumor biology • Allows testing of immunotherapeutic approaches	• Lack of efficient communication between mouse immune components and the human ones • More technically challenging than conventional PDXs • Limited source of HSCs • Limited window of opportunity to work with PBMC-engrafted models

[1] As humanized immune system mouse models have not yet been applied extensively to cancer, this table extrapolates the advantages and disadvantages of their use in personalized medicine to their application in cancer studies.

intravenous delivery of chemotherapeutics is difficult to perform. Similarly, complex multidrug combinations used in the clinic are seldom studied in orthotopic or GEMM models of metastasis, and synthetic toxicities are rarely uncovered before reaching clinical trials in humans. Nevertheless, when well designed and executed, studies including mouse models provide invaluable information and are essential for the development of novel targeted therapies. As patient stratification increases, medicine becomes more personalized: greater knowledge of the mutation spectrum of the patient, coupled with an increased number of such targeted therapies, leads to more appropriate and effective tailored treatments.

In our view, the appropriate experimental design to preclinically validate new therapeutic targets or agents will likely require the use of multiple mouse models to adequately address the various aspects of both mechanism of action and efficacy.

References

Abaan OD, Polley EC, Davis SR, Zhu YJ, Bilke S, Walker RL, et al. (2013). The exomes of the NCI-60 panel: a genomic resource for cancer biology and systems pharmacology. *Cancer Res*, **73**, 4372–4382.

Agliano A, Martin-Padura I, Mancuso P, Marighetti P, Rabascio C, Pruneri G, et al. (2008). Human acute leukemia cells injected in NOD/LtSz-scid/IL-2Rgamma null mice generate a faster and more efficient disease compared to other NOD/scid-related strains. *Int J Cancer*, **123**, 2222–2227.

Annibali D, Whitfield JR, Favuzzi E, Jauset T, Serrano E, Cuartas I, et al. (2014). Myc inhibition is effective against glioma and reveals a role for Myc in proficient mitosis. *Nat Commun*, **5**, 4632.

Bassuk AG, and Leiden JM. (1995). A direct physical association between ETS and AP-1 transcription factors in normal human T cells. *Immunity*, **3**, 223–237.

Bentley NJ, Eisen T, and Goding CR. (1994). Melanocyte-specific expression of the human tyrosinase promoter: activation by the microphthalmia gene product and role of the initiator. *Mol Cell Biol*, **14**, 7996–8006.

Boughey JC, Yu J, Yin P, Gao B, Sinnwell JP, Moyer AM, Visscher DW, Conners AL, Dockter TJ, Kalari KR, Jones KN, McLaughlin SA, Copland

JA, Moreno-Aspitia A, Northfelt DW, Gray RJ, Suman VJ, Weinshilboum RM, Goetz MP and Wang L. (2014). Patient-derived xenografts from breast cancer patients before and after neoadjuvant chemotherapy: a prospective study. *J Clin Oncol*, **32**, 5s abstr 11130.

Bertotti A, Migliardi G, Galimi F, Sassi F, Torti D, Isella C, et al. (2011). A molecularly annotated platform of patient-derived xenografts ('xenopatients') identifies HER2 as an effective therapeutic target in cetuximab-resistant colorectal cancer. *Cancer Discovery*, **1**, 508–523.

Brehm MA, Cuthbert A, Yang C, Miller DM, DiIorio P, Laning J, et al. (2010). Parameters for establishing humanized mouse models to study human immunity: analysis of human hematopoietic stem cell engraftment in three immunodeficient strains of mice bearing the IL2rgamma(null) mutation. *Clin Immunol*, **135**, 84–98.

Brehm MA, Jouvet N, Greiner DL, and Shultz LD. (2013). Humanized mice for the study of infectious diseases. *Curr Opin Immunol*, **25**, 428–435.

Cancer Genome Atlas Research Network, Weinstein JN, Collisson EA, Mills GB, Shaw KRM, Ozenberger BA, et al. (2013). The Cancer Genome Atlas Pan-Cancer analysis project. *Nat Genet*, **45**, 1113–1120.

Chen Z, Cheng K, Walton Z, Wang Y, Ebi H, Shimamura T, et al. (2012). A murine lung cancer co-clinical trial identifies genetic modifiers of therapeutic response. *Nature*, **483**, 613–617.

Chin L, Hahn WC, Getz G, and Meyerson M. (2011). Making sense of cancer genomic data. *Genes Dev*, **25**, 534–555.

Choi SYC, Lin D, Gout PW, Collins CC, Xu Y, and Wang Y. (2014). Lessons from patient-derived xenografts for better in vitro modeling of human cancer. *Adv Drug Deliv Rev*, **79-80**, 222–237.

Daniel VC, Marchionni L, Hierman JS, Rhodes JT, Devereux WL, Rudin CM, et al. (2009). A primary xenograft model of small-cell lung cancer reveals irreversible changes in gene expression imposed by culture in vitro. *Cancer Res*, **69**, 3364–3373.

Dankort D, Curley DP, Cartlidge RA, Nelson B, Karnezis AN, Damsky WE, et al. (2009). Braf(V600E) cooperates with Pten loss to induce metastatic melanoma. *Nat Genet*, **41**, 544–552.

Danner R, Chaudhari SN, Rosenberger J, Surls J, Richie TL, Brumeanu T-D, et al. (2011). Expression of HLA class II molecules in humanized NOD.Rag1KO.IL2RgcKO mice is critical for development and function of human T and B cells. *PLoS One*, **6**, e19826.

De Giovanni C, Nicoletti G, Landuzzi L, Romani F, Croci S, Palladini A, et al. (2012). Human responses against HER-2-positive cancer cells in human immune system-engrafted mice. *Br J Cancer*, **107**, 1302–1309.

DeNardo DG, Johansson M, and Coussens LM. (2008). Immune cells as mediators of solid tumor metastasis. *Cancer Metastasis Rev*, **27**, 11–18.

Denayer T, Stöhr T, and Van Roy M. (2014). Animal models in translational medicine: validation and prediction. *New Horiz Transl Med*, **2**, 5–11.

Denton PW, and Garcia JV. (2011). Humanized mouse models of HIV infection. *AIDS Rev*, **13**, 135–148.

DeRose YS, Wang G, Lin Y-C, Bernard PS, Buys SS, Ebbert MTW, et al. (2011). Tumor grafts derived from women with breast cancer authentically reflect tumor pathology, growth, metastasis and disease outcomes. *Nat Med*, **17**, 1514–1520.

Diede SJ, Yao Z, Keyes CC, Tyler AE, Dey J, Hackett CS, et al. (2013). Fundamental differences in promoter CpG island DNA hypermethylation between human cancer and genetically engineered mouse models of cancer. *Epigenetics*, **8**, 1254–1260.

Doroshow JH, and Kummar S. (2014). Translational research in oncology—10 years of progress and future prospects. *Nat Rev Clin Oncol*, **11**, 649–662.

Douglas ML, Boucaut KJ, Antalis TM, Higgins C, Pera MF, Stuttgen MA, et al. (2001). An orthotopic xenograft model of human nonseminomatous germ cell tumour. *Br J Cancer*, **85**, 608–611.

Fichtner I, Rolff J, Soong R, Hoffmann J, Hammer S, Sommer A, et al. (2008). Establishment of patient-derived non-small cell lung cancer xenografts as models for the identification of predictive biomarkers. *Clin Cancer Res*, **14**, 6456–6468.

Fiebig HH, Neumann HA, Henss H, Koch H, Kaiser D, and Arnold H. (1985). Development of three human small cell lung cancer models in nude mice. *Recent Results Cancer Res*, **97**, 77–86.

Frese KK, Neesse A, Cook N, Bapiro TE, Lolkema MP, Jodrell DI, et al. (2012). nab-Paclitaxel potentiates gemcitabine activity by reducing cytidine deaminase levels in a mouse model of pancreatic cancer. *Cancer Discovery*, **2**, 260–269.

Gajewski TF, Schreiber H, and Fu Y-X. (2013). Innate and adaptive immune cells in the tumor microenvironment. *Nat Immunol*, **14**, 1014–1022.

Giovanella BC, Stehlin JW, Wall ME, Wani MC, Nicholas AW, Liu LF, Sibert R and Potmesil M (1989). DNA Topoisomerase I-targeted chemotherapy of human colon colorectal cancer in xenografts. *Science*, **246**, 1046–1048.

Gossen M, Freundlieb S, Bender G, Muller G, Hillen W, and Bujard H. (1995). Transcriptional activation by tetracyclines in mammalian cells. *Science*, **268**, 1766–1769.

Guerra C, and Barbacid M. (2013). Genetically engineered mouse models of pancreatic adenocarcinoma. *Mol Oncol*, **7**, 232–247.

Haluska FG, Tsao H, Wu H, Haluska FS, Lazar A, and Goel V. (2006). Genetic alterations in signaling pathways in melanoma. *Clin Cancer Res*, **12**, 2301s–2307s.

Hanahan D, and Coussens LM. (2012). Accessories to the crime: functions of cells recruited to the tumor microenvironment. *Cancer Cell*, **21**, 309–322.

Hanahan D, and Weinberg RA. (2011). Hallmarks of cancer: the next generation. *Cell*, **144**, 646–674.

Henderson VC, Kimmelman J, Fergusson D, Grimshaw JM, and Hackam DG. (2013). Threats to validity in the design and conduct of preclinical efficacy studies: a systematic review of guidelines for in vivo animal experiments. *PLoS Med*, **10**, e1001489.

Herter-Sprie GS, Kung AL, and Wong K-K. (2013). New cast for a new era: preclinical cancer drug development revisited. *J Clin Invest*, **123**, 3639–3645.

Hesami P, Holzapfel BM, Taubenberger A, Roudier M, Fazli L, Sieh S, et al. (2014). A humanized tissue-engineered in vivo model to dissect interactions between human prostate cancer cells and human bone. *Clin Exp Metastasis*, **31**, 435–446.

Hidalgo M, Amant F, Biankin AV, Budinska E, Byrne AT, Caldas C, et al. (2014). Patient-derived xenograft models: an emerging platform for translational cancer research. *Cancer Discovery*, **4**, 998–1013.

Hidalgo M, Bruckheimer E, Rajeshkumar NV, Garrido-Laguna I, De Oliveira E, Rubio-Viqueira B, et al. (2011). A pilot clinical study of treatment guided by personalized tumorgrafts in patients with advanced cancer. *Mol Cancer Ther*, **10**, 1311–1316.

Hingorani SR, Petricoin EF, Maitra A, Rajapakse V, King C, Jacobetz MA, et al. (2003). Preinvasive and invasive ductal pancreatic cancer and its early detection in the mouse. *Cancer Cell*, **4**, 437–450.

Hingorani SR, Wang L, Multani AS, Combs C, Deramaudt TB, Hruban RH, et al. (2005). Trp53R172H and KrasG12D cooperate to promote chromosomal instability and widely metastatic pancreatic ductal adenocarcinoma in mice. *Cancer Cell*, **7**, 469–483.

Hiramatsu H, Nishikomori R, Heike T, Ito M, Kobayashi K, Katamura K, et al. (2003). Complete reconstitution of human lymphocytes from cord blood CD34+ cells using the NOD/SCID/gammacnull mice model. *Blood*, **102**, 873–880.

Hoff Von DD, Ervin T, Arena FP, Chiorean EG, Infante J, Moore M, et al. (2013). Increased survival in pancreatic cancer with nab-paclitaxel plus gemcitabine. *N Engl J Med*, **369**, 1691–1703.

Hollingshead MG, Stockwin LH, Alcoser SY, Newton DL, Orsburn BC, Bonomi CA, et al. (2014). Gene expression profiling of 49 human tumor xenografts from in vitro culture through multiple in vivo passages–strategies for data mining in support of therapeutic studies. *BMC Genomics*, **15**, 393.

Holzapfel BM, Thibaudeau L, Hesami P, Taubenberger A, Holzapfel NP, Mayer-Wagner S, et al. (2013). Humanised xenograft models of bone metastasis revisited: novel insights into species-specific mechanisms of cancer cell osteotropism. *Cancer Metastasis Rev*, **32**, 129–145.

Hooijmans CR, de Vries R, Leenaars M, Curfs J, and Ritskes-Hoitinga M. (2011). Improving planning, design, reporting and scientific quality of animal experiments by using the Gold Standard Publication Checklist, in addition to the ARRIVE guidelines. *Br J Pharmacol*, **162**, 1259–1260.

Houghton JA, and Taylor DM. (1978). Growth characteristics of human colorectal tumours during serial passage in immune-deprived mice. *Br J Cancer*, **37**, 213–223.

Houghton JA, Houghton PJ, and Green AA. (1982). Chemotherapy of childhood rhabdomyosarcomas growing as xenografts in immune-deprived mice. *Cancer Res*, **42**, 535–539.

Hutchinson L, and Kirk R. (2011). High drug attrition rates-where are we going wrong? *Nat Rev Clin Oncol*, **8**, 189–190.

Iida N, Dzutsev A, Stewart CA, Smith L, Bouladoux N, Weingarten RA, et al. (2013). Commensal bacteria control cancer response to therapy by modulating the tumor microenvironment. *Science*, **342**, 967–970.

Inman KS, Francis AA, and Murray NR. (2014). Complex role for the immune system in initiation and progression of pancreatic cancer. *World J Gastroenterol*, **20**, 11160–11181.

Ishikawa F, Yasukawa M, Lyons B, Yoshida S, Miyamoto T, Yoshimoto G, et al. (2005). Development of functional human blood and immune systems in NOD/SCID/IL2 receptor {gamma}chain(null) mice. *Blood*, **106**, 1565–1573.

Ito A, Ishida T, Yano H, Inagaki A, Suzuki S, Sato F, et al. (2009). Defucosylated anti-CCR4 monoclonal antibody exercises potent ADCC-mediated antitumor effect in the novel tumor-bearing humanized NOD/Shi-scid, IL-2Rgamma(null) mouse model. *Cancer Immunol Immunother*, **58**, 1195–1206.

Ito M, Hiramatsu H, Kobayashi K, Suzue K, Kawahata M, Hioki K, et al. (2002). NOD/SCID/gamma(c)(null) mouse: an excellent recipient mouse model for engraftment of human cells. *Blood*, **100**, 3175–3182.

Ito R, Takahashi T, Katano I, and Ito M. (2012). Current advances in humanized mouse models. *Cell Mol Immunol*, **9**, 208–214.

Jackson EL. (2001). Analysis of lung tumor initiation and progression using conditional expression of oncogenic K-ras. *Genes Dev*, **15**, 3243–3248.

Janne PA, Shaw AT, Pereira JR, Jeannin G, Vansteenkiste J, Barrios C, et al. (2013). Selumetinib plus docetaxel for KRAS-mutant advanced non-small-cell lung cancer: a randomised, multicentre, placebo-controlled, phase 2 study. *Lancet Oncol*, **14**, 38–47.

Ji H, Wang Z, Perera SA, Li D, Liang M-C, Zaghlul S, et al. (2007). Mutations in BRAF and KRAS converge on activation of the mitogen-activated protein kinase pathway in lung cancer mouse models. *Cancer Res*, **67**, 4933–4939.

Jin K, Teng L, Shen Y, He K, Xu Z, and Li G. (2010). Patient-derived human tumour tissue xenografts in immunodeficient mice: a systematic review. *Clin Transl Oncol*, **12**, 473–480.

Jones S, Hruban RH, Kamiyama M, Borges M, Zhang X, Parsons DW, et al. (2009). Exomic sequencing identifies PALB2 as a pancreatic cancer susceptibility gene. *Science*, **324**, 217–217.

Kalluri R, and Zeisberg M. (2006). Fibroblasts in cancer. *Nat Rev Cancer*, **6**, 392–401.

Kaufmann SHE. (2008). Immunology's foundation: the 100-year anniversary of the Nobel Prize to Paul Ehrlich and Elie Metchnikoff. *Nat Immunol*, **9**, 705–712.

Kauppila S, Stenbäck F, Risteli J, Jukkola A, and Risteli L. (1998). Aberrant type I and type III collagen gene expression in human breast cancer in vivo. *J Pathol*, **186**, 262–268.

Kurtz JE and Ray-Coquard I. (2012). PI3 kinase inhibitors in the clinic: an update. *Anticancer Res*, **32**(7), 2463–2470.

Kwon M-C, and Berns A. (2013). Mouse models for lung cancer. *Mol Oncol*, **7**, 165–177.

Levental KR, Yu H, Kass L, Lakins JN, Egeblad M, Erler JT, et al. (2009). Matrix crosslinking forces tumor progression by enhancing integrin signaling. *Cell*, **139**, 891–906.

Li H, Yang AL, Chung YT, Zhang W, Liao J, and Yang G-Y. (2013). Sulindac inhibits pancreatic carcinogenesis in LSL-KrasG12D-LSL-Trp53R172H-Pdx-1-Cre mice via suppressing aldo-keto reductase family 1B10 (AKR1B10). *Carcinogenesis*, **34**, 2090–2098.

Lin WM, Baker AC, Beroukhim R, Winckler W, Feng W, Marmion JM, et al. (2008). Modeling genomic diversity and tumor dependency in malignant melanoma. *Cancer Res*, **68**, 664–673.

Littlewood TD, Hancock DC, Danielian PS, Parker MG, and Evan GI. (1995). A modified oestrogen receptor ligand-binding domain as an improved switch for the regulation of heterologous proteins. *Nucleic Acids Res*, **23**, 1686–1690.

Liu P, Cheng H, Roberts TM, and Zhao JJ. (2009). Targeting the phosphoinositide 3-kinase pathway in cancer. *Nat Rev Drug Discovery*, **8**, 627–644.

Liu P, Cheng H, Santiago S, Raeder M, Zhang F, Isabella A, et al. (2011). Oncogenic PIK3CA-driven mammary tumors frequently recur via PI3K pathway-dependent and PI3K pathway-independent mechanisms. *Nat Med*, **17**, 1116–1120.

Loukopoulos P, Kanetaka K, Takamura M, Shibata T, Sakamoto M, and Hirohashi S. (2004). Orthotopic transplantation models of pancreatic adenocarcinoma derived from cell lines and primary tumors and displaying varying metastatic activity. *Pancreas*, **29**, 193–203.

Makkouk A, and Weiner GJ. (2015). Cancer immunotherapy and breaking immune tolerance: new approaches to an old challenge. *Cancer Res*, **75**, 5–10.

Malaney P, Nicosia SV, and Davé V. (2014). One mouse, one patient paradigm: new avatars of personalized cancer therapy. *Cancer Lett*, **344**, 1–12.

Massó-Vallés D, Jauset T, Serrano E, Sodir NM, Pedersen K, Affara NI, Whitfield JR, Beaulieu ME, Evan GI, Elias L, Arribas J and Soucek L. (2015). Ibrutinib exerts potent anti-fibrotic and anti-tumor activities in mouse models of pancreatic adenocarcinoma. *Cancer Res*, **75**(8), 1675–1681.

Melkus MW, Estes JD, Padgett-Thomas A, Gatlin J, Denton PW, Othieno FA, et al. (2006). Humanized mice mount specific adaptive and innate immune responses to EBV and TSST-1. *Nat Med*, **12**, 1316–1322.

Milsom CC, Lee CR, Hackl C, Man S, and Kerbel RS. (2013). Differential post-surgical metastasis and survival in SCID, NOD-SCID and NOD-

SCID-IL-2Rγ(null) mice with parental and subline variants of human breast cancer: implications for host defense mechanisms regulating metastasis. *PLoS One*, **8**, e71270.

Moro M, Bertolini G, Tortoreto M, Pastorino U, Sozzi G, and Roz L. (2012). Patient-derived xenografts of non small cell lung cancer: resurgence of an old model for investigation of modern concepts of tailored therapy and cancer stem cells. *J Biomed Biotechnol*, **2012**, 568567.

Morton CL, and Houghton PJ. (2007). Establishment of human tumor xenografts in immunodeficient mice. *Nat Protoc*, **2**, 247–250.

Némati F, Sastre-Garau X, Laurent C, Couturier J, Mariani P, Desjardins L, et al. (2010). Establishment and characterization of a panel of human uveal melanoma xenografts derived from primary and/or metastatic tumors. *Clin Cancer Res*, **16**, 2352–2362.

Pearson T, Greiner DL, and Shultz LD. (2008). Creation of 'humanized' mice to study human immunity. *Curr Protoc Immunol*, **Chapter 15**, Unit 15.21.

Picard D, Salser SJ, and Yamamoto KR. (1988). A movable and regulable inactivation function within the steroid binding domain of the glucocorticoid receptor. *Cell*, **54**, 1073–1080.

Platt RJ, Chen S, Zhou Y, Yim MJ, Swiech L, Kempton HR, et al. (2014). CRISPR-Cas9 knockin mice for genome editing and cancer modeling. *Cell*, **159**, 440–455.

Posch C, Moslehi H, Feeney L, Green GA, Ebaee A, Feichtenschlager V, et al. (2013). Combined targeting of MEK and PI3K/mTOR effector pathways is necessary to effectively inhibit NRAS mutant melanoma in vitro and in vivo. *Proc Natl Acad Sci U S A*, **110**, 4015–4020.

Provenzano PP, Eliceiri KW, Campbell JM, Inman DR, White JG, and Keely PJ. (2006). Collagen reorganization at the tumor-stromal interface facilitates local invasion. *BMC Med*, **4**, 38.

Qian B-Z, and Pollard JW. (2010). Macrophage diversity enhances tumor progression and metastasis. *Cell*, **141**, 39–51.

Regales L, Balak MN, Gong Y, Politi K, Sawai A, Le C, et al. (2007). Development of new mouse lung tumor models expressing EGFR T790M mutants associated with clinical resistance to kinase inhibitors. *PLoS One*, **2**, e810.

Remer M, Al-Shamkhani A, Glennie M, and Johnson P. (2014). Mogamulizumab and the treatment of CCR4-positive T-cell lymphomas. *Immunotherapy*, **6**, 1187–1206.

Reyal F, Guyader C, Decraene C, Lucchesi C, Auger N, Assayag F, et al. (2012). Molecular profiling of patient-derived breast cancer xenografts. *Breast Cancer Res*, **14**, R11.

Rongvaux A, Willinger T, Martinek J, Strowig T, Gearty SV, Teichmann LL, et al. (2014). Development and function of human innate immune cells in a humanized mouse model. *Nat Biotechnol*, **32**, 364–372.

Rubinson DA, Hochster HS, Ryan DP, Wolpin BM, McCleary NJ, Abrams TA, et al. (2014). Multi-drug inhibition of the HER pathway in metastatic colorectal cancer: results of a phase I study of pertuzumab plus cetuximab in cetuximab-refractory patients. *Invest New Drugs*, **32**, 113–122.

Rubio-Viqueira B, and Hidalgo M. (2009). Direct in vivo xenograft tumor model for predicting chemotherapeutic drug response in cancer patients. *Clin Pharmacol Ther*, **85**, 217–221.

Rubio-Viqueira B, Jimeno A, Cusatis G, Zhang X, Iacobuzio-Donahue C, Karikari C, et al. (2006). An in vivo platform for translational drug development in pancreatic cancer. *Clin Cancer Res*, **12**, 4652–4661.

Ryan DP, Hong TS, and Bardeesy N. (2014). Pancreatic adenocarcinoma. *N Engl J Med*, **371**, 2140–2141.

Saunders TL. (2011). Inducible transgenic mouse models. *Methods Mol Biol*, **693**, 103–115.

Sánchez-Rivera FJ, Papagiannakopoulos T, Romero R, Tammela T, Bauer MR, Bhutkar A, et al. (2014). Rapid modelling of cooperating genetic events in cancer through somatic genome editing. *Nature*, **516**, 428–431.

Scannell JW, Blanckley A, Boldon H and Warrington B. (2012). Diagnosing the decline in pharmaceutical R&D efficiency. *Nat Rev Drug Discovery*, **11**, 191–200.

Seitz G, Pfeiffer M, Fuchs J, Warmann SW, Leuschner I, Vokuhl C, et al. (2010). Establishment of a rhabdomyosarcoma xenograft model in human-adapted mice. *Oncol Rep*, **24**, 1067–1072.

Shultz LD, Saito Y, Najima Y, Tanaka S, Ochi T, Tomizawa M, et al. (2010). Generation of functional human T-cell subsets with HLA-restricted immune responses in HLA class I expressing NOD/SCID/IL2r gamma(null) humanized mice. *Proc Natl Acad Sci U S A*, **107**, 13022–13027.

Singh M, Lima A, Molina R, Hamilton P, Clermont AC, Devasthali V, et al. (2010). Assessing therapeutic responses in Kras mutant cancers using genetically engineered mouse models. *Nat Biotechnol*, **28**, 585–593.

Singh M, Murriel CL, and Johnson L. (2012). Genetically engineered mouse models: closing the gap between preclinical data and trial outcomes. *Cancer Res*, **72**, 2695–2700.

Sivanand S, Peña-Llopis S, Zhao H, Kucejova B, Spence P, Pavia-Jimenez A, et al. (2012). A validated tumorgraft model reveals activity of dovitinib against renal cell carcinoma. *Sci Transl Med*, **4**, 137ra75.

Snook AE, and Waldman SA. (2013). Advances in cancer immunotherapy. *Discov Med*, **15**, 120–125.

Soucek L, Whitfield J, Martins CP, Finch AJ, Murphy DJ, Sodir NM, et al. (2008). Modelling Myc inhibition as a cancer therapy. *Nature*, **455**, 679–683.

Soucek L, Whitfield JR, Sodir NM, Masso-Valles D, Serrano E, Karnezis AN, et al. (2013). Inhibition of Myc family proteins eradicates KRas-driven lung cancer in mice. *Genes Dev*, **27**, 504–513.

Stellas D, Szabolcs M, Koul S, Li Z, Polyzos A, Anagnostopoulos C, et al. (2014). Therapeutic effects of an anti-Myc drug on mouse pancreatic cancer. *J Natl Cancer Inst*, **106**, dju320.

Subklewe M, Geiger C, Lichtenegger FS, Javorovic M, Kvalheim G, Schendel DJ, et al. (2014). New generation dendritic cell vaccine for immunotherapy of acute myeloid leukemia. *Cancer Immunol Immunother*, **63**, 1093–1103.

Sugamura K, Asao H, Kondo M, Tanaka N, Ishii N, Ohbo K, et al. (1996). The interleukin-2 receptor gamma chain: its role in the multiple cytokine receptor complexes and T cell development in XSCID. *Annu Rev Immunol*, **14**, 179–205.

Takenaka K, Prasolava TK, Wang JCY, Mortin-Toth SM, Khalouei S, Gan OI, et al. (2007). Polymorphism in Sirpa modulates engraftment of human hematopoietic stem cells. *Nat Immunol*, **8**, 1313–1323.

Tentler JJ, Tan A-C, Weekes CD, Jimeno A, Leong S, Pitts TM, et al. (2012). Patient-derived tumour xenografts as models for oncology drug development. *Nat Rev Clin Oncol*, **9**, 338–350.

Thibaudeau L, Taubenberger AV, Holzapfel BM, Quent VM, Fuehrmann T, Hesami P, et al. (2014). A tissue-engineered humanized xenograft model of human breast cancer metastasis to bone. *Dis Model Mech*, **7**, 299–309.

Traggiai E, Chicha L, Mazzucchelli L, Bronz L, Piffaretti J-C, Lanzavecchia A, et al. (2004). Development of a human adaptive immune system in cord blood cell-transplanted mice. *Science*, **304**, 104–107.

van Rijn RS, Simonetti ER, Hagenbeek A, Hogenes MCH, de Weger RA, Canninga-van Dijk MR, et al. (2003). A new xenograft model for graft-versus-host disease by intravenous transfer of human peripheral blood mononuclear cells in RAG2-/- gammac-/- double-mutant mice. *Blood*, **102**, 2522–2531.

Vogt PK, Kang S, and Elsliger MA. (2007). Cancer-specific mutations in phosphatidylinositol 3-kinase. *Trends Biochem Sci*, **32**, 342–349.

Wang J, Miletic H, Sakariassen PØ, Huszthy PC, Jacobsen H, Brekkå N, et al. (2009). A reproducible brain tumour model established from human glioblastoma biopsies. *BMC Cancer*, **9**, 465.

Wang J, Xia T-S, Liu X-A, Ding Q, Du Q, Yin H, et al. (2010). A novel orthotopic and metastatic mouse model of breast cancer in human mammary microenvironment. *Breast Cancer Res Treat*, **120**, 337–344.

Wege AK, Ernst W, Eckl J, Frankenberger B, Vollmann-Zwerenz A, Männel DN, et al. (2011). Humanized tumor mice-a new model to study and manipulate the immune response in advanced cancer therapy. *Int J Cancer*, **129**, 2194–2206.

Werner-Klein M, Proske J, Werno C, Schneider K, Hofmann H-S, Rack B, et al. (2014). Immune humanization of immunodeficient mice using diagnostic bone marrow aspirates from carcinoma patients. *PLoS One*, **9**, e97860.

Whitfield JR, Littlewood T, Evan GI, Soucek L (2015). The estrogen receptor fusion system in mouse models: a reversible switch. *Cold Spring Harb Protoc*, **2015**(3), 227–234.

Williams SA, Anderson WC, Santaguida MT, and Dylla SJ. (2013). Patient-derived xenografts, the cancer stem cell paradigm, and cancer pathobiology in the 21st century. *Lab Invest*, **93**, 970–982.

Willinger T, Rongvaux A, Strowig T, Manz MG, and Flavell RA. (2011). Improving human hemato-lymphoid-system mice by cytokine knock-in gene replacement. *Trends Immunol*, **32**, 321–327.

Wilson CH, Gamper I, Perfetto A, Auw J, Littlewood TD, and Evan GI. (2014). The kinetics of ER fusion protein activation in vivo. *Oncogene*, **33**, 4877–4880.

Woiterski J, Ebinger M, Witte KE, Goecke B, Heininger V, Philippek M, et al. (2013). Engraftment of low numbers of pediatric acute lymphoid and myeloid leukemias into NOD/SCID/IL2Rcγnull mice reflects individual leukemogenecity and highly correlates with clinical outcome. *Int J Cancer*, **133**, 1547–1556.

Yahata T, Ando K, Nakamura Y, Ueyama Y, Shimamura K, Tamaoki N, et al. (2002). Functional human T lymphocyte development from cord blood CD34+ cells in nonobese diabetic/Shi-scid, IL-2 receptor gamma null mice. *J Immunol*, **169**, 204–209.

Yang X, Zhang X, Mortenson ED, Radkevich-Brown O, Wang Y, and Fu Y-X. (2013). Cetuximab-mediated tumor regression depends on innate and adaptive immune responses. *Mol Ther*, **21**, 91–100.

Yoshimura M, Endo S, Hioki K, Ueyama Y, and Ohnishi Y. (1997). Chemotherapeutic profiles of human tumors implanted in SCID mice showing appreciable inconsistencies with those in nude mice. *Exp Anim*, **46**, 153–156.

Zadeh-Khorasani M, Nolte T, Mueller TD, Pechlivanis M, Rueff F, Wollenberg A, et al. (2013). NOD-scid IL2R γnull mice engrafted with human peripheral blood mononuclear cells as a model to test therapeutics targeting human signaling pathways. *J Transl Med*, **11**, 4.

Zhu GG, Risteli L, Mäkinen M, Risteli J, Kauppila A, and Stenbäck F. (1995). Immunohistochemical study of type I collagen and type I pN-collagen in benign and malignant ovarian neoplasms. *Cancer*, **75**, 1010–1017.

Zvibel I, Smets F, and Soriano H. (2002). Anoikis: roadblock to cell transplantation? *Cell Transplant*, **11**, 621–630.

Chapter 7

Tumor Microenvironment, Therapeutic Resistance, and Personalized Medicine

Yu Sun
Key Laboratory of Stem Cell Biology, Institute of Health Sciences, Shanghai Institutes for Biological Sciences, Chinese Academy of Sciences and Shanghai Jiao Tong University School of Medicine, Shanghai 200031, China
Collaborative Innovation Center of Systems Biomedicine, Shanghai Jiao Tong University School of Medicine, Shanghai 200240, China
VA Seattle Medical Center, Seattle, WA 98108, USA
Department of Medicine, University of Washington, Seattle, WA 98195, USA
sunyu@sibs.ac.cn

Tumors do not behave as a homogeneous mixture of transformed cells but as a complex multicellular entity wherein malignant cells actively communicate with adjacent benign tissue elements that compose the stroma. The interaction between cancer cells and the stroma—which usually comprises genetically stable cells and the extracellular matrix (ECM)—together constitute a "tumor microenvironment" (TME), a paradigm now considered one of the typical hallmarks of cancer [1]. Indeed, the first pathological description of TME as a disease-relevant term was delivered by a "seed and soil" concept initially proposed by Paget in the late 19th

Cancer Genetics and Genomics for Personalized Medicine
Edited by Il-Jin Kim

ISBN 978-981-4669-87-0 (Hardcover), 978-1-315-36476-6 (eBook)
www.panstanford.com

century to delineate the distinct patterns of recurrent metastatic sites of human mammary cancer, and to plausibly interpret the tropism of tumor metastases to specific sites [2]. In health bodies, the diverse compartments of the microenvironment do not stay as quiet bystanders, but significantly regulate tumor initiation, disease progression, metastatic development, and more importantly, therapeutic response. Among multiple TME-implicated activities, clinical response to therapies is the major factor that directly determines the long-term fate of patients who undergo anticancer interventions.

7.1 TME Orchestrates Disease Progression and Dominates Therapeutic Response

As a most lethal age-related pathology that imperils human health, cancer co-evolves with the surrounding TME to achieve successful outgrowth and ensuing metastasis that is responsible for the majority of cancer-related death [3]. Although continued progress has been made in therapeutic methodologies, disease relapse with diminished response remains a significant challenge and confers poor prognosis in clinical oncology. Cancer resistance involves intrinsic mechanisms that are determined by pre-existing genetic and/or epigenetic properties of malignant cells, including enhanced drug efflux, blunted apoptotic signaling, increased metabolic activities, loss of specific oncogenes, gain of stem cell plasticity, and strengthened DNA damage repair machinery, all fueled by mutation-selective pressure that engenders clonal expansion and create tumor heterogeneity [4, 5]. In contrast, extrinsic resistance of cancer cells driven by the TME represents a seemingly minor but essentially pivotal modality that substantially influences therapeutic efficacy.

First, the TME mediates innate resistance prior to cytotoxic treatment events, which is through regular mutual interactions between cancer cells and the neighboring TME components. This force differs from inherent resistance, which is based on original alterations at the genomic and/or epigenomic levels of cancer cells. Second, acquired resistance conferred by the TME usually emerges as a host adaptive response to pharmacological insults. Specifically, TME-provoked resistance generates profound impacts to local disease foci and shapes cancer evolution trajectory under varying treatment pressures in clinical settings.

7.1.1 Cancer-Associated Fibroblasts

In solid tissues, fibroblasts constitute the structural framework and maintain physiological homeostasis as a predominant mesenchymal lineage. However, cancer-associated fibroblasts (CAFs) are functionally distinct from their normal counterparts and frequently demonstrate pathological relevance. In the microenvironment milieu, normal fibroblasts can be transformed into CAFs once stimulated by local tissue-derived proteins such as fibroblast growth factor (FGF), monocyte chemotactic protein 1 (MCP-1), platelet-derived growth factor (PDGF), tissue inhibitor of metalloproteinase 1 (TIMP-1) and tumor transforming growth factor β (TGF-β) [6, 7]. Despite the tumor-suppressive capacity in certain malignancies including pancreatic ductal adenocarcinoma (PDA), CAFs exhibit aggressive proliferation, augmented ECM deposition, enhanced cytokine synthesis/secretion (for instance, fibroblast growth factor 7 (FGF7); hepatocyte growth factor (HGF); interleukin 6 (IL-6); PDGF; stromal cell-derived factor 1, (SDF-1); and vascular endothelial growth factor (VEGF)), a unique stromal phenotype characterized with a chemoresistance-triggering secretome that can be abolished upon mTOR/4E-BP1 translation pathway blockade [4, 8].

Following activation in the TME, CAFs generate proinflammatory factors that promote tumor progression in a NF-κB-dependent manner, drive leukocyte infiltration, stimulating angiogenesis and vascular permeability [9–11]. Primary tumor select for bone metastatic seeds in the TME based on the interaction between Src pathway- activated cancer cells and chemokine C-X-C motif ligand 12 (CXCL12)/insulin-like growth factor 1 (IGF1)-secreting CAFs, indicating the evolution of metastatic traits in a primary foci and the distant metastasis [12].

7.1.2 Vasculature System

The tumor vascular network is derived from new vessels, through co-option and modification of mature vessels, or via differentiation of endothelial precursors from bone marrow, each contributing to vascular development and heterogeneity [4]. Vessel formation involves remodeling of pre-existing vascular basement membranes, and the pattern varies depending on the tissue type. Although a functional vasculature is vital for both tumor survival and metastatic progression by supplying oxygen and nutrients, poorly organized tumor vasculatures cause emergence of hypoxia and limited growth factor feeding. Co-operation of several cell types in the TME, including endothelial cells, pericytes, and bone marrow-derived precursor cells, is fundamental for tumor vascularization, although such synergism is often modulated by hypoxia [13, 14].

Spatial interval from vasculatures to tumor foci generates an infiltration gradient associated with drug distribution to cancer cells within the tissue, while microvessel density (MVD) is a significant prognostic factor for clinical outcome in malignancies including breast, liver, lung, and lower lip squamous cell carcinoma (LLSCC) [15–18]. Mesenchymal stem cells (MSCs), tumor-associated macrophages (TAMs), and CAFs, all contribute to tumor vascularization by secreting a variety of angiogenesis-related ligands into the TME. Particularly, increased expression of the pro-angiogenic factor VEGFA is correlated with worse prognosis in metastatic colorectal, lung, and renal cell cancers [19].

7.1.3 Extracellular Matrix

The ECM is produced by multiple TME cell types and weaves an intricate fiber network not only providing structural support but regulating cellular activities [20]. In early life stages the ECM prevents cancer initiation as a safeguard, while at a later stage it actively increases pathological incidence, particularly tumorigenesis [21]. TME-associated ECM essentially differs from that of the normal tissue, serving as a basic scaffold for cancer cell invasion driven by chemotaxis [22]. Interplay between cancer cells and ECM elements is dynamic and goes far beyond spatial contact. In breast cancer, malignant cell attachment to ECM alters their polarization and causes resistance to etoposide-induced apoptosis [23]. Specifically, cell adhesion-mediated drug resistance (CAM-DR) depends on association of integrin to ECM components, including fibronectin, collagen, and laminin [5]. Integrin-implicated cancer resistance is not limited to chemotherapies, but applies to radiation and targeted therapies that attenuate activities of receptor tyrosine kinase (RTK), including the epidermal growth factor receptor (EGFR) [24, 25]. Enhanced ECM biosynthesis and elevated collagen fiber crosslinking in the reactive stroma further enables cancer progression through augmenting integrin signaling [26]. Of note, desmoplastic stroma is a physical barrier for chemotherapeutic agent delivery and influences tumor vascular architecture [27].

According to the ECM composition, patients including those with breast neoplasia can be stratified into several classes and predicted with distinct clinical outcomes [28]. Tumors with increased protease inhibitor expression in the ECM are associated with optimistic prognosis, but those with upregulated integrins and matrix metallopeptidases (MMPs) correlate with poor prognosis with recurrence possibility. To this end, a bioinformatic approach to predict the in silico "matrisome," defined as the ensemble of ECM proteins and associated factors, was developed, which revealed that primary foci of diverse metastatic potential differ in their composition of both tumor- and stroma-derived ECM components [29]. Therefore, targeting the extracellular microenvironment that is structurally adjacent and functionally linked to tumors may provide an extra option for therapeutic intervention.

7.1.4 Immune Cells

In healthy individuals, an intact immune system functionally mounts antitumor responses, but mechanisms of immune suppression under pathological conditions can prevent this process and promote disease progression. Both positive and negative checkpoint signals are involved in T-cell activation as host response to prevent damage and minimize autoimmunity. Cytotoxic T cells in the TME are activated through continuous engagement of inhibitory cell surface receptors, including cytotoxic T lymphocyte–associated antigen 4

(CTLA-4) and programmed death 1 (PD-1) receptor, through ligand overexpression [30]. Alternatively, antitumor T cell responses are suppressed within an immunosuppressive milieu, as tumorigenesis induces expansion of a myeloid-derived suppressor cell (MDSC) population that comprises early myeloid progenitors, immature dendritic cells (DCs), neutrophils, and monocytes, a process that implicates endocrine communication with the immune system through enhanced secretion of chemokines, including granulocyte–macrophage colony-stimulating factor (GM-CSF) and granulocyte colony-stimulating factor (G-CSF) [31, 32]. Tumor-induced recruitment of immunosuppressive myeloid lineages not only diminishes adaptive immunity, but fosters angiogenesis through production of basic fibroblast growth factor (bFGF), TGF-β and VEGFA [33]. MDSCs also inhibit effector T-cell proliferation, activation, and migration, while enhancing immunosuppressive regulatory T-cell expansion [34].

In addition to myeloid cells, B cells can advance tumor progression by enhancing chronic immune response, inflammatory mast cells are involved in tumorigenesis and angiogenesis, while TAMs are functionally plastic and can adjust their polarization to motivate cancer cells and interfere with innate immunity [4]. Data from TME analysis indicates that angiogenesis, hypoxia, and high TAM content relative to cytotoxic T-cell number correlates with a worse outcome of breast tumors, although the overall stromal signature derived from patient biopsies predicts outcome independently of tumor subtypes [35]. A recent study demonstrated that TAMs also mediate resistance to gemcitabine by overexpression of cytidine deaminase, which inactivate gemcitabine in pancreatic ductal cancer cells [36]. Increased secretion of IL-10 by TAMs is responsible for drug resistance in breast cancer, a case associated with elevated bcl-2 expression and upregulation of STAT3 signaling, while neutralizing antibody to TAM-derived IL-10 can reverse such resistance [37]. Further, TAMs produce IL-6, which allows expansion of CD44(+) cancer stem cells (CSCs) in hepatocellular carcinoma (HCC) and correlates with recurrence after therapy [38]. Blockade of IL6 signaling with tocilizumab, a drug approved by the Food and Drug Administration (FDA) for rheumatoid arthritis, effectively inhibited TAM-stimulated resistance of CD44(+) cells.

7.1.5 TME-Derived Exosomes

Exosomes originate in large multivescular bodies (MVBs), form in the extracellular milieu upon MVB fusion with the plasma membrane, share the same topology as a cell, and harbor a large pool of biologically active components [39]. In the TME, exosomes are secreted by diverse cell types and circulate with body fluid in forms of blood, urine, saliva, ascitic fluid, and amniotic fluid [40].

Despite the presence of typical paracrine mechanisms that play critical roles in cancer-stroma communication, exosome shedding has recently emerged as another mode of TME-implicated pathologies. For example, melanoma-released exosomes increased both growth and metastasis of primary tumors and expansion of bone marrow–derived cells (BMDCs) toward a pro-metastatic phenotype through the receptor tyrosine kinase MET-mediated signaling, effects significantly impaired by inhibition of exosomes [41]. Exosome of cancer cell origin can functionally trap therapeutic antibodies such as rituximab and trastuzumab, eventually compromising immunotherapy efficacy [42, 43]. Fibroblast-secreted exosomes enhance protrusion and motility of breast cancer cells through Wnt-planar cell polarity (PCP) signaling, while co-implantation of cancer cells and fibroblasts promotes metastasis that is associated with PCP activities and Cd81 in fibroblast exosomes [44]. Noncoding RNAs and transposable elements in stromal exosomes trigger retinoic acid–inducible gene 1 (RIG-I), a pattern recognition receptor, to activate signal transducer and activator of transcription 1 (STAT1)-associated antiviral signaling, engage NOTCH3 pathway, and expand breast cancer (BCa) subpopulations adept at countering treatments and acquiring tumor-initiating ability [45]. Although soluble factors in natural killer (NK) cell–derived exosomes induce tumor cytotoxicity and activate immune cells, they do not disrupt resting immune cells, indicating special targeting of activated cells by these exosomes [46]. Dendritic cell (DC)-released exosomes, termed dexosomes, are in clinical trials to evaluate their potential as a cancer vaccine, an investigation encouraged by the fact that dexosome immunization induces CD8+ cytotoxic T cells and inhibits tumor growth [47].

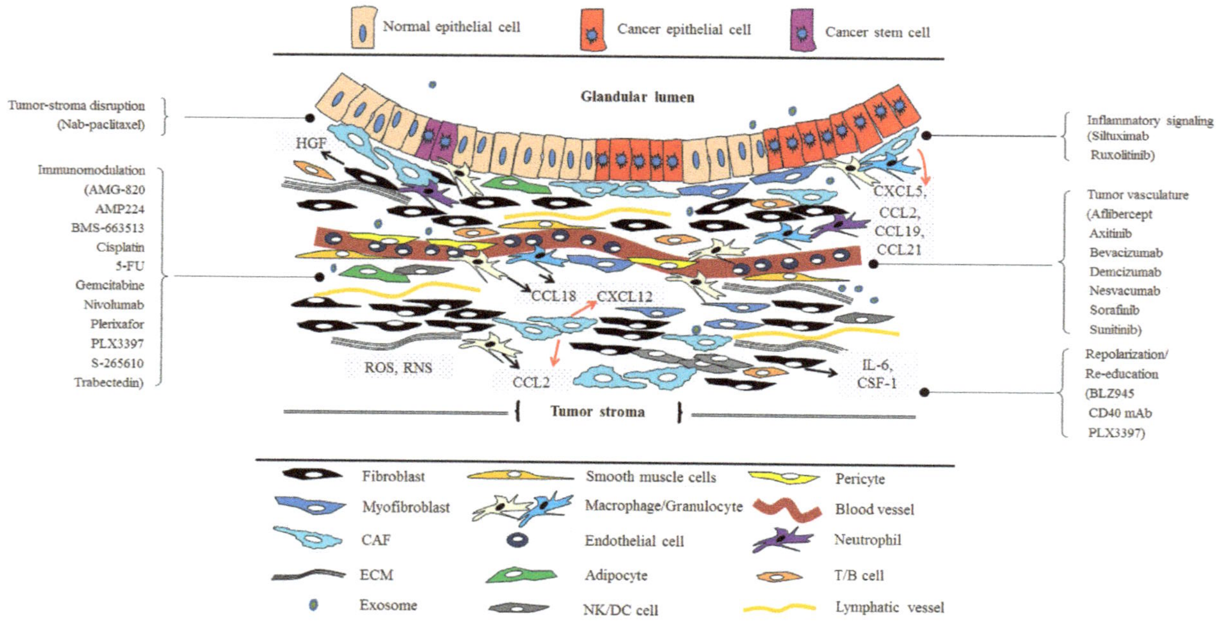

Figure 7.1 Schematic outline of the TME including diverse structural and functional components that are targetable by therapeutic agents. A typical TME consists of many cell lineages, most of which are directly or indirectly correlated with accelerated disease progression and altered treatment response, thereby receiving significant attention in recent years. For example, CAFs are activated upon tumorigenesis, generating a myriad of growth factors, cytokines and chemokines, acquiring

Figure 7.1 (*Contd.*) a proinflammatory phenotype and composing a major source of soluble factors that promote neoplastic cell survival and dissemination. Immune cells in the TME display remarkable plasticity, particularly TAMs and MDSCs form an immunosuppressive and tumor-permissive phenotype that extends to multiple T and B cells. The tumor vasculature evolves from either initiating angiogenesis or mature vessels, supplying oxygen and nutrients to allow both tumor growth and metastatic progression. Inflammatory pathways activated in the TME further spurs release of soluble materials including several classic tumor-associated interleukins including IL-6 and IL-8. The ECM constituents including fibronectin, collagen and laminin can cause cell adhesion-mediated drug resistance (CAM-DR), depending on the status of their association with integrin. Both stromal cells in the TME and cancer cells secrete exosomes that function in a paracrine or autocrine manner to enhance tumor-induced immune suppression, angiogenesis and form premetastatic niches. The majority of the TME components are being evaluated for their individual relevance in pathological contribution to tumor development, and clinically subject to chemotherapies and/or targeted therapies. Pharmacological agents are organized into specific groups that manipulate certain aspects of the TME, with a handful of FDA-approved and those in preclinical or clinical trials randomly selected as representative entries. *Abbreviations*: CAF, cancer-associated fibroblast; ECM, extracellular matrix; TAM, tumor-associated macrophage; MDSC, myeloid-derived suppressor cell; NK, natural killer cell; DC, dendritic cell; HGF, hepatocyte growth factor; CCL, chemokine C-C motif ligand; CXCL, chemokine C-X-C motif ligand; CSF, colony-stimulating factor; IL-6, interleukin 6; ROS, reactive oxygen species, RNS, reactive nitrogen species; 5-FU, 5-Fluorouracil. The specific targets of exemplary targeted agents are as follows: AMG820 (CSF-1R), AMP224 (PD1), BMS-663513 (CD137), Nivolumab (PD1 or OX40), Plerixafor (CXCR4), PLX3397 (CSF-1R), S-265610 (CXCR2), Siltuximab (IL-6), Ruxolitinib (JAK1/2), Aflibercept (decoy receptor), Axitinib (VEGFRs, PDGFRs, KIT), Bevacizumab (VEGFA), Demcizumab (DLL4), Nesvacumab (ANG2), Sorafenib (VEGFRs, RAF, PDGFRs, KIT), Sunitinib (VEGFRs, PDGFRs, FLT3, CSF-1R), BLZ945 (CSF-1R), CD40 mAb (CD40), PLX3397 (CSF-1R). Agents not given corresponding targets belong to the class of chemotherapeutic drugs. Abbreviations for above targets refer to Table 7.1 of this chapter [4, 21].

Overall, the functional relevance of TME in altering tumor phenotypes and shaping disease evolution is supported by rising lines of experimental and clinical evidence, further highlighting the intricate profiles of TME structure and function that entered the spotlight of targeted therapies (Fig. 7.1). Although there are reports regarding the positive role of local microenvironments in sensitizing tumors to treatments [48, 49], recent studies revealed the capability of TME in distorting outcomes of nonconventional interventions, including targeted regimens. Kinase inhibitors to BRAF, ALK, or EGFR induce a complicated network composed of secreted factors in human and mouse melanoma and lung cancer [50]. The therapy-induced secretome (TIS) promotes dissemination and metastasis of drug-resistance clones and contribute to incomplete disease regression, but tumors are susceptible to combination agents that inhibit both BRAF and PI3K/Akt/mTOR intracellular signaling pathways. Thus, the long-term understanding of anticancer therapies that plausibly target only malignant cell populations should be well updated with the rapidly augmenting discoveries in cancer biology and advances in medical oncology.

7.2 Treatment-Activated TME Confers Acquired Resistance and Creates Barriers to a Clinical Cure

7.2.1 Damage Responses of the TME Offset Therapy-Enforced Tumor Regression

First introduced into clinics in 1940s for preliminary uses of nitrogen mustards and antifolate drugs, chemotherapy constitutes the mainstay of modern anticancer regimes and demonstrated remarkable potency in controlling malignancies including leukemia, lymphoma and most solid tumors [51]. However, the technical limitations of chemotherapy discovered by the early researchers still remain, and the major downside of such an action mode is due to its low efficiency in distinguishing between normal and cancer cells. Exposure of patients to chemotherapy or radiation, a genotoxic approach sharing similar treatment principles in oncology, inevitably triggers in vivo damage responses that significantly modify therapeutic outcomes [52].

Cellular senescence occurs among normal cells in response to oncogenic overexpression (oncogene-induced senescence, OIS) or telomeric attrition (replicative senescence, RS), serving as a physiological checkpoint to cause cell cycle arrest and prevent carcinogenesis. Importantly, senescent cells also exhibit another visage characterized with production of a plethora of cytokines, chemokines, proteases, and growth factors, a signature termed the senescence-associated secretory phenotype (SASP) or the senescence messaging secretome (SMS) [53, 54]. Despite the cell-autonomous function of certain SASP factors such as IL-6 and IL-8 in reinforcing the senescence state, diverse SASP effectors perform as cell non-autonomous molecules that degrade basement membrane, promote cancer resistance, enhance metastatic potential, together accelerating tumorigenic progression [55, 56].

Genotoxicity delivered by chemotherapy and radiation not only destroy the mass of tumors, but attack benign compartments in the TME. Treatment response correlates with activation of a DNA damage secretory program (DDSP) in stromal cells, as is intensively recognized by recent studies. Besides typical senescence attributes that are consistent with SASP features, the TME in prostate, breast, and ovarian cancer patients displayed remarkable upregulation of WNT16B upon neoadjuvant or classical chemotherapy, while NF-κB appeared as a key signaling node that actively mediates WNT16B biosynthesis [57]. Cell culture assays and tumor xenograft models demonstrated the protective effect of stroma-derived WNT16B, indicating that WNT16B produced by the TME attenuates cancer cell apoptosis caused by genotoxicity, and confers tumor insensitivity to therapeutic agents through activation of a stroma-intensive secretory phenotype. Phenotypes driven by the DDSP are distinct and the associated secretion is robust, with many extracellular proteins overexpressed to 10-fold or even higher (for instance, MMP1 76-fold and WNT16B 34-fold), and the effects are supposed to be chronical as evidenced by effector secretion in TME niches of clinical specimens collected even months (or years) after radical surgery [5, 57].

In parallel, soluble factors from the TME can modulate survival of lymphoma cells following administration of genotoxic drugs. IL-6 and Timp-1 from endothelial cells of the thymus create a chemoresistance niche, promoting the survival and expansion of cancer cells remaining in the local microenvironments as minimal residual disease (MRD), and priming for eventual tumor relapse [58]. Therefore, resident benign cells in the TME respond to genotoxic stress by activating a paracrine secretory program, which can be co-opted by malignant cells to survive frontline chemotherapy. Although such a response is named acute stress–associated phenotype (ASAP) as distinct from SASP which upon DNA damage develops gradually within 7 to 10 days after establishment of senescence markers, the process is essentially covered by the concept of DDSP, a more general term that can be exploited to adjust therapeutic settings such as fine-tuning of metronomic chemotherapy [59].

In contrast to organs that comprise mostly well-differentiated cell lineages, the bone marrow is a significant source enriched with both stem cells and varying progenies beyond myeloid components mobilized and recruited to the TME in response to therapies. Upon cisplatin-exerted damage, mesenchymal stem cells (MSCs) in the bone marrow secrete polyunsaturated fatty acids, which promote resistance to various agents and even act at a systemic level [60]. Administration of antiangiogenic agents, including doxorubicin, 5-fluorouracil, and paclitaxel, causes the bone marrow to generate circulating endothelial cells that are homed to tumor sites [61]. Interestingly, antibodies that target VEGF receptor 2 (VEGFR2) or SDF-1α abolished such recruitment and enhanced treatment efficacy, a case yet not reproducible by gemcitabine-involved combination therapy. Further, a protective niche was recently identified within the bone marrow, created by cancer propagating cells (CPCs) of leukemia following cytarabine and daunorubicin treatment [62]. The "refuge" niche evolves during therapeutic intervention, while recruited mesenchymal cells secreted chemokine C–C motif ligand 3 (CCL3) and growth differentiation factor 15 (GDF15) that actively built a therapy-induced shield, transitioned from Nestin+ cells to alpha smooth muscle actin (a-SMA)+ cells under the influence of TGF-β and finally turned into fiber residues. Similarly, a lymphatic niche is present in melanoma and permits CD133+ tumor initiating cell (TIC) survival, accumulation, and metastasis after dacarbazine administration, while radiation-damaged TME generates a stem cell niche and confers resistance to glioma cells that reside preferentially in specific sites [63, 64].

7.2.2 Modified Differentiation and Immune Responses in the TME Decrease Therapeutic Outcomes

Polarized epithelial cells can convert to motile mesenchymal cells before in vivo relocation, a process termed the epithelial to mesenchymal transition (EMT). Although frequently delineating cellular plasticity in tissue homeostasis and organ developments, EMT represents a typical cell-intrinsic mechanism that allows development of cancer resistance, and studies have proved that damage-induced expression of soluble factors in the TME can promote such a phenotypic switch of cancer cells to affect therapeutic outcome [53, 57]. Both altered microRNA expression and abnormal intracellular signaling are linked with EMT, and a "signature" of 30 experimentally validated microRNAs and genes is recently reported that images the crosstalk between paracrine TGF-β, Notch, and Wnt released from TME resident cells during EMT and reinforce their regulation of epithelial plasticity in tumors [65].

Interestingly, melanoma cells can de-differentiate in a mouse model upon adoptive cell transfer (ACT), resulting in melanocytic antigen loss and resistance to ensuing ACT cycles, a phenomenon mediated by tumor necrosis factor alpha (TNF-a) released by tumor-infiltrating T cells and macrophages in the TME [66]. Several TME-released cytokines, including TNF-a, mediate mesenchymal differentiation through NF-kB in glioblastoma, eventually causing radioresistance [67]. In addition, macrophages can be programed by chemotherapies including platinum analogs, which induce an M1-M2 transition in the TME, while low-dose irradiation promoting an iNOS+/M1 phenotype that orchestrate T-cell immunotherapy toward an immunosuppressive state allows the recruitment of cytotoxic T cells in the tumor [68, 69]. Hypoxia-primed breast cancer cells chemoattract and polarize macrophages to pro-angiogenic M2-polarized subtype via eotaxin and oncostatin M [70]. However, blockade of eotaxin/oncostatin M not only prevented emergence of the M2-polarized phenotype and retarded tumor progression in 4T1/BALB/c-syngenic mouse model of breast cancer but enhanced the efficacy of the antiangiogenic agent bevacizumab, implying therapies targeting these cytokines are optimal for tumors that are refractory to antiangiogenic drugs.

Collectively, damage responses of resident lineages in the TME, altered differential status of cancer cells, and rewired immune cell signaling significantly compromise therapeutic efficacy. Successfully unraveling the complexity and exploiting the mechanisms represents an optimal avenue to minimize resistance acquired from a treatment-remodeled TME. To date, it is increasingly evident that durable responses will indeed optimally result from novel regimens that both target cancer cells and manipulate the TME components. To this end, development of targeted agents to control the impact of stromal cell types is now blooming and ongoing clinical trials allow one to envisage a clear scenario (Table 7.1).

7.3 Overcoming Challenges of Personalized Cancer Therapy Requires Translation of Biological Insights into the Clinic

7.3.1 Implications of Personalized Cancer Therapy in an Era of Precision Medicine

Personalized cancer therapy (PCT) takes advantage of informative clues from the tumor and its microenvironment, together with distinct conditions of the patient, to tailor therapeutic regimes and treat the disease more effectively with less toxicity. PCT delivers a similar concept with "individualized cancer therapy" (ICT) that aims to design strategies for a person diagnosed with cancer, through covering drug sensitivity testing, cancer biomarkers and bioinformatics detection, pharmacogenetics, individualized antimetastatic therapy, drug combinations, assistant chemotherapy, cost-effectiveness consideration, and guidelines for utilization/reimbursement of molecular diagnosis.

Table 7.1 A representative panel of therapeutic agents that target specific compartments of the TME, a critical part of pathological etiology that calls for attention

Molecule	Target	Molecular type	Company	Status
ECM/fibroblasts				
Sonidegib	SMO	Small molecule	Novartis	Phase II (NCT01708174, NCT01757327, NCT02195973)
Vasculature				
Bevacizumab	VEGFA	Antibody	Genentech/Roche	FDA-approved ((BLA) 125085)
Vandetanib	VEGFRs, PDGFRs, EGFR	Small molecule	AstraZeneca	FDA-approved ((NDA) 022405)
Sunitinib	VEGFRs, PDGFRs, FLT3, CSF1R	Small molecule	Pfizer	FDA-approved ((NDA) 021938)
Axitinib	VEGFRs, PDGFRs, KIT	Small molecule	Pfizer	FDA-approved ((NDA) 022324)
Sorafenib	VEGFRs, RAF, PDGFRs, KIT	Small molecule	Bayer	FDA-approved ((NDA) 021923)
Pazopanib	VEGFRs, PDGFRs, KIT	Small molecule	GlaxoSmithKline	FDA-approved ((NDA) 022465)
Cabozantinib	VEGFR2, RET, MET	Small molecule	Exelixis	FDA-approved ((NDA) 023756)
Ziv-aflibercept	VEGFA, VEGFB, PIGF	Receptor-Fc fusion	Regeneron	FDA-approved ((BLA) 125418)
AMG 386	ANG2	RP-Fc fusion protein	Amgen	Phase III (NCT01204749, NCT01493505, NCT01281254)
Parsatuzumab	EGFL-7	Antibody	Genentech/Roche	Phase II (NCT01399684, NCT01366131)
Enoticumab	DLL4	Antibody	Regeneron	Phase I (NCT00871559)
Demcizumab	DLL4	Antibody	OncoMed	Phase I (NCT00744562, NCT01189968, NCT01189942, NCT01189929)

(*Contd.*)

Table 7.1 (*Contd.*)

Molecule	Target	Molecular type	Company	Status
Nesvacumab	ANG2	Antibody	Regeneron	Phase I (NCT01688960, NCT01271972)
Immune				
Ipilimumab	CTLA-4	Antibody	Bristol-Myers Squibb	FDA-approved ((BLA) 125377)
Sipuleucel-T	PAP	DC vaccine	Dendreon	FDA-approved ((BLA) 125197)
Aldesleukin	IL-2	RP	Prometheus	FDA-approved ((BLA) 103293)
IFNα-2b	IFN-α receptor	RP	Merck	FDA-approved ((BLA) 103132)
MK-3475	PD1	Antibody	Merck	Phase III (NCT01866319)
Nivolumab	PD1	Antibody	Bristol-Myers Squibb	Phase III (NCT01642004, NCT01668784, NCT01673867, NCT01721746, NCT01721772, NCT01844505)
Nivolumab	OX40	Antibody	Bristol-Myers Squibb and PPMC	Phase III (NCT01642004, NCT01668784, NCT01673867, NCT01721746, NCT01721772, NCT01844505)
MPDL3280A	PDL1	Antibody	Genentech/Roche	Phase II (NCT01846416)
PLX3397	KIT, CSF1R, FLT3	Small molecule	Plexxikon	Phase II (NCT01349036)
BMS-663513	CD137 (4-1BB)	Antibody	Bristol-Myers Squibb	Phase II (NCT00612664)
Blinatumomab	CD3 and CD19	Bi-specific scFv	Amgen	Phase II (NCT01741792, NCT01466179, NCT01207388, NCT01471782, NCT00560794, NCT01209286)

AMG 820	CSF1R	Antibody	Amgen	Phase I (NCT01444404)
AMP-224	PD1	Antibody	GlaxoSmithKline	Phase I (NCT01352884)
TRX-518	GITR	Antibody	GITR, Inc.	Phase I (NCT01239134)
IMC-CS4	CSR1R	Antibody	ImClone/Eli Lilly	Phase I (NCT01346358)
CP-870,893	CD40	Antibody	Pfizer	Phase I (NCT00711191, NCT01008527, NCT00607048, NCT01456585, NCT01103635)

References listed in the status column pertain to the molecule as a TME-modifying agent, either the FDA application, where approved, or the national clinical trial identification of the oncology trial in the latest phase is listed (note that in some cases the drug may also be tested or approved for an indication for which it acts directly on the tumor cell compartment, which will not be referenced here). ANG2, angiopoietin 2; BLA, biological license application; CSF1R, colony stimulating factor 1 receptor; CTLA-4, cytotoxic T-lymphocyte-associated antigen 4; DC, dendritic cell; DLL4, delta-like 4; ECM, extracellular matrix; EGFL-7, epidermal growth factor like 7; EGFR, epidermal growth factor receptor; FDA, Food and Drug Administration; FLT3, Fms-like tyrosine kinase 3; GBM, glioblastoma multiforme; GITR, glucocorticoid-induced TNFR-related; IFN, interferon; IL-2, interleukin 2; MMP, matrixmetalloproteinase; NCT, national clinical trial; NDA, new drug application; PAP, prostatic acid phosphatase; PD-1, programmed death-1; PDAC, pancreatic ductal adenocarcinoma; PDGFR, Platelet-derived growth factor receptor; PDL1, programmed death ligand 1; PPMC, Portland Providence Medical Center; RP, recombinant peptide; scFv, single-chain Fv; SMO, smoothened; VEGF, vascular endothelial growth factor; VEGFR, vascular endothelial growth factor receptor. Table adapted from Junttila and de Sauvage [4], with permission from *Nature*, copyright 2013.
Note, agents that either failed to be effective in clinical trials or have been officially terminated are removed from the current list.

A major obstacle to controlling cancer progression in patients is the inappropriate selection and administration of anticancer agents. Researchers and clinicians are now beginning to focus on PCT, to improve therapeutic quality and outcomes by prescribing the most appropriate and effective drugs. Using genetic, molecular, and bioinformatics data and modern experimental techniques, known oncogenes can be detected more easily than ever before. A good case in point is that the average speed of genetic sequencing has increased substantially since the completion of Human Genome Project (HGP).

However, the promise of PCT cannot be realized if we fail to promptly identify and incorporate information from biomarkers that are predictive of therapeutic response and drug resistance into an individualized level, thereby improving the interventional efficacy and preventing emergence of cancer resistance. This way, clinicians can avoid applying adjuvant therapies to patients at low risk of recurrence and stop administrating with inefficient agents. To date there are excellent biomarkers in predicting response and benefits in clinics, as illustrated by the case of BRAF inhibitor vemurafenib, which gives significantly higher response rates relative to dacarbazine, a standard care for patients with metastatic melanoma [71]. Such striking results inspired analysis of our approach to the classification and treatment of malignancies in an age of molecular markers and targeted therapy [72].

Despite the exciting progress in establishing such "fine-tuning" approaches, there are numerous challenges ahead to overcome before successful implementation of PCT in clinical oncology. Given the biological complexity and dynamic properties of the TME, technical obstacles such as limitations of molecular tests, lack of appropriate pharmaceuticals, and shortcomings of reimbursement mechanisms, the "hype" surrounding PCT has yet to be adjusted with realistic methodologies [73]. Unfortunately, most of currently available treatment regimens offer prolonged survival for only a small handful of cancer patients, and how to break such a stalemate through designing innovative, tactful, and precise technologies is a critical step toward major advances in clinical oncology.

As a previously unrecognized pathological implication of TME, the concrete traits of niches surrounding the tumor foci influence the way biomarkers for PCT are developed and utilized for cancer patients. In general, TME-specific markers for clinical detection should be established by differentiating responses between cancer cells and the TME components upon a given therapy. So far, the majority of therapy-oriented biomarkers adapted for PCT are present in subtypes of distinct cancers and involve activated oncogenes targetable by small molecule inhibitors such as ALK or EGFR mutations in lung carcinomas and HER2 or BRCA1/2 overexpression in breast tumors, or by monoclonal antibodies such as KRAS in colorectal cancers and BRAF in melanomas [50, 74–76]. Indeed, a substantial percentage of these patients do not respond to mainstream treatments, and high-resolution genomic assays failed to interpret the discordance, inferring significant TME implication in reducing therapeutic efficacy in clinics.

7.3.2 Significance of Preclinical Studies in Promoting PCT Advancement

In contrast to clinical investigation, preclinical models can effectively incorporate tumor–stroma interactions and help identify candidate markers for medical validation. Although it is unlikely for the experimental data from preclinical studies to perfectly recapitulate clinical conditions, the resulting improvement can prominently fill the gap between benchside findings and bedside etiology.

For decades, cell lines have represented a powerful tool for cancer research, as they facilitate studies of genetic drivers or suppressors of human malignancies and multiple markers associated with response to therapeutics, as supported by the systematic translation from Cancer Cell Line Encyclopedia (CCLE) [77, 78]. Further, since primary cancer cells without long-term exposure to ex vivo conditions remain higher TME-dependence than established cell lines, it is reasonable to expand the research of patient-derived cancer cells by in vitro co-culture assays and/or orthotopic animal models [79]. In contrast to conventional two-dimensional (2D) culture, the three-dimensional (3D) organoid system that integrates updated technologies into new tissue models for biomarker discovery offers valuable opportunities to identify new drug targets and predict patient responses to anticancer treatments [80]. Furthermore, data contributed by orthotopic studies involving patient samples from distinct pathological phases help disclose specific tumor–microenvironment interaction patterns, a subdiscipline of cancer biology that received significant attention in recent years particular those engaging humanized mouse models.

Despite the genomic difference between human and mouse, sufficient overlapping properties provide unique opportunities to generate informative clues across multiple pathophysiological barriers. Taking advantage of such similarity, humanized mouse models are being developed, which can effectively reconstitute malignancies with intact immunity in mice carrying tumors of human origin. For instance, simultaneous implantation of human peripheral blood lymphocytes (PBL), DC, and prostate tumors to NOD/SCID/IL-2Rγnull (NSG) mice builds in a huPBL-NSG model that recapitulates the human tumor-infiltrating lymphocyte (TIL) development and allows evaluation of tumor and immune system interaction, which can be performed in neither human patients nor regular mice [81]. More importantly, the capacity to manipulate special TME cell lineages or functional components permits hypothesis examination and outcome assessment in a humanized system that is more relevant than conventional mouse models, including the first generation of patient-derived xenografts (PDXs). Following such an exemplary concept, human hematopoietic stem and progenitor cells (HSPCs) are expanded to reconstitute the radiation-depleted bone marrow of a NSG mouse which received a patient's tumor, generating a "XactMice" [82]. Immune and stromal cells of human origin propagated in these animals help recapitulate the TME of an implanted xenograft, circumvent any potential genetic drift, and provide an optimal tumor model to guide patient treatment, significantly accelerating PCT studies.

Novel animal models that simulate tumor progression in a clinically relevant TME represent a cornerstone for development of anticancer pharmacology. The design, discovery, and exploitation of chemotherapeutic and targeted agents require further understanding the pathological roles of the TME. Furthermore, prospective outputs from continued PCT studies will allow tailoring of therapies on a patient-to-patient mode, a process not determined by cancer cells exclusively but rather through combining tumor features and TME biomarkers, the latter experimentally probed with high-throughput techniques. Treatment response evaluated by such orthogonal approaches holds significant promise as the next frontier that leads the translational medicine and empowers innovation of clinical oncology.

7.4 Concluding Remarks and Future Outlooks

Next-generation sequencing (NGS) enables genome-wide personalized oncology efforts with the specialty and infrastructure necessary for identification and prioritization of tumor genome variants, as piloted by the Oncomine Comprehensive Panel (OCP), which represents a streamlined and broadly applicable targeted NGS system to advance precision oncology [83]. In contrast, biological mechanisms of resistance to conventional cytotoxic chemotherapeutics and targeted therapies designed for specific molecules share multiple features, including activated prosurvival pathways and disabled apoptotic machineries, together constituting response variation that awaits precise treatments [84]. Pathological roles of the TME in modulating tumor sensitivity is being increasingly recognized as a key aspect for development of modern anticancer therapeutics. It is well accepted that tumor–stroma interplay dictates cell signaling, metabolism, survival, proliferation, and therapeutic response. Resistance or sensitization of malignancies to certain pharmacological agents depend not only on intrinsic traits of cancer cells, but substantially on the context of a specific TME. Given the mounting arsenal of anticancer agents, improving preclinical models, and the advent of state-of-the-art screening technologies, unprecedented opportunities are available to monitor and overcome drug resistance. For example, through clinical assessment of predictive biomarkers to enable patient stratification according to the TME status which images distinct TME responses particularly upon exposure of cancer patients to genotoxic chemotherapies, rational therapeutic combinations can be considered to minimize the impact of an activated TME on disease exacerbation (Fig. 7.2) [5, 57, 85, 86]. Further, advances in in vitro screening platforms and in vivo orthotopic xenograft mouse models

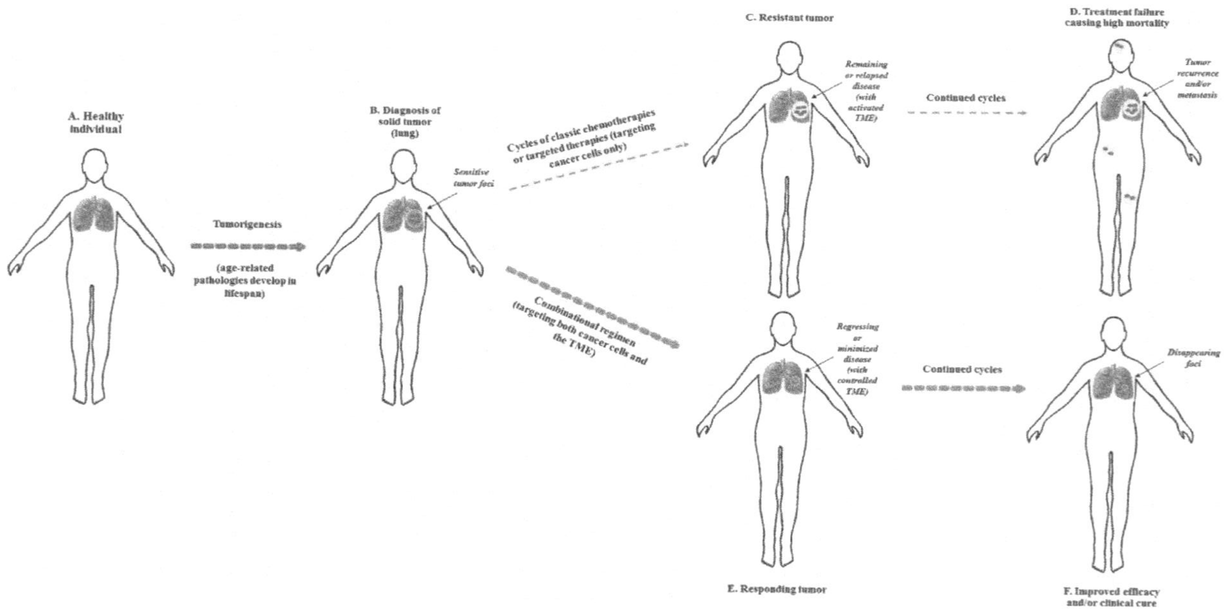

Figure 7.2 Cancer resistance acquired from a therapeutically activated TME poses major challenges to clinical oncology. (A) The normal human individual without malignancies at all. (B) In the course of aging, certain neoplasia develops such as the lung

Figure 7.2 (*Contd.*) adenocarcinoma. (C) Under conventional treatments including regular chemotherapies and targeted therapies, the tumor responds by shrinking in volume at the early stage, but subsequent development of resistant phenotype prevents further success. (D) With continued cycles of treatments, disease remains and relapses with high frequency of metastasis to distant organs, causing clinical failure and high incidence of cancer-related death. (E) In contrast, novel therapies that combine cancer-targeting drugs and TME-oriented agents significantly enhance the sensitivity of tumors, and disease responds well. (F) After further consolidation with combined therapies, the tumor enters rapid regression with optimal efficacy, the disease can be ideally cured with remarkably improved clinical outcome. In both cancer biology and medical oncology, thoroughly understanding the mechanisms of tumor-stroma interaction is critical for discovery of new therapeutic options, as proved by the ways through which major breakthroughs were made in past decades.

have enabled comprehensive characterization of the implication of the TME in modifying therapeutic efficacy. There is no doubt that recent advances will ideally bridge the gap between preclinical studies and clinical trials of anticancer regimens.

The clinical goal of PCT is to target aberrations that drive tumor growth and survival, by administering the appropriate drug combination for each individual patient. This is becoming increasingly achievable with rapid progress in cutting-edge technologies to characterize tumors and the expanding repertoire of therapeutic agents that target TME constituents. However, there remain numerous challenges to take and surpass before we can conquer the diseases on the promise of PCT. In the current era of precision medicine, partnerships between academia and industry as well as significant philanthropic support are essential to facilitate comprehensive molecular characterization to demonstrate that it benefits patients [73]. However, ultimately PCT will become the both clinically preferred and financially available model, by treating a cancer patient with the right therapy, at the first time, generating best responses and providing medical cures.

References

1. Hanahan, D., and R. A. Weinberg, Hallmarks of cancer: the next generation. *Cell,* 2011. **144**(5), 646–674.
2. Paget, S., The distribution of secondary growths in cancer of the breast. *Lancet,* 1889. **133**(3421), 571–573.
3. Chen, F., et al., New horizons in tumor microenvironment biology: challenges and opportunities. *BMC Med,* 2015. **13**(1), 278.
4. Junttila, M. R., and F. J. de Sauvage, Influence of tumour microenvironment heterogeneity on therapeutic response. *Nature,* 2013. **501**(7467), 346–354.
5. Sun, Y., Translational horizons in the tumor microenvironment: harnessing breakthroughs and targeting cures. *Med Res Rev,* 2015. **35**(2), 408–436.
6. Quail, D. F., and J. A. Joyce, Microenvironmental regulation of tumor progression and metastasis. *Nat Med,* 2013. **19**(11), 1423–1437.

7. Song, T., et al., TIMP-1 activated carcinoma-associated fibroblasts inhibit tumor apoptosis by activating SDF1/CXCR4 signaling in hepatocellular carcinoma. *Oncotarget*, 2015. **6**(14), 12061–12079.
8. Duluc, C., et al., Pharmacological targeting of the protein synthesis mTOR/4E-BP1 pathway in cancer-associated fibroblasts abrogates pancreatic tumour chemoresistance. *EMBO Mol Med*, 2015. **7**(6), 735–753.
9. Erez, N., et al., Cancer-associated fibroblasts are activated in incipient neoplasia to orchestrate tumor-promoting inflammation in an NF-kappaB-dependent manner. *Cancer Cell*, 2010. **17**(2), 135–147.
10. Erez, N., et al., Cancer associated fibroblasts express pro-inflammatory factors in human breast and ovarian tumors. *Biochem Biophys Res Commun*, 2013. **437**(3), 397–402.
11. Nagasaki, T., et al., Interleukin-6 released by colon cancer-associated fibroblasts is critical for tumour angiogenesis: anti-interleukin-6 receptor antibody suppressed angiogenesis and inhibited tumour-stroma interaction. *Br J Cancer*, 2014. **110**(2), 469–478.
12. Zhang, X. H., et al., Selection of bone metastasis seeds by mesenchymal signals in the primary tumor stroma. *Cell*, 2013. **154**(5), 1060–1073.
13. Weis, S. M., and D. A. Cheresh, Tumor angiogenesis: molecular pathways and therapeutic targets. *Nat Med*, 2011. **17**(11), 1359–1370.
14. Semenza, G. L., Cancer-stromal cell interactions mediated by hypoxia-inducible factors promote angiogenesis, lymphangiogenesis, and metastasis. *Oncogene*, 2013. **32**(35), 4057–4063.
15. Tredan, O., et al., Drug resistance and the solid tumor microenvironment. *J Natl Cancer Inst*, 2007. **99**, 1441–1454.
16. Yuan, M. M., et al., TLR3 expression correlates with apoptosis, proliferation and angiogenesis in hepatocellular carcinoma and predicts prognosis. *BMC Cancer*, 2015. **15**(1), 245.
17. Ariotti, C., et al., VEGFR1 and VEGFR2 in lip carcinogenesis and its association with microvessel density. *Tumour Biol*, 2015. **36**(9), 7285–7292.
18. Lee, J. G., and R. Wu, Erlotinib-cisplatin combination inhibits growth and angiogenesis through c-MYC and HIF-1 alpha in EGFR-mutated lung cancer in vitro and in vivo. *Neoplasia*, 2015. **17**(2), 190–200.
19. Hegde, P. S., et al., Predictive impact of circulating vascular endothelial growth factor in four phase III trials evaluating bevacizumab. *Clin Cancer Res*, 2013. **19**(4), 929–937.

20. Martino, M. M., et al., Growth factors engineered for super-affinity to the extracellular matrix enhance tissue healing. *Science*, 2014. **343**(6173), 885–888.
21. Klemm, F., and J. A. Joyce, Microenvironmental regulation of therapeutic response in cancer. *Trends Cell Biol*, 2015. **25**(4), 198–213.
22. Friedl, P., and S. Alexander, Cancer invasion and the microenvironment: plasticity and reciprocity. *Cell,* 2011. **147**(5), 992–1009.
23. Weaver, V. M., et al., beta4 integrin-dependent formation of polarized three-dimensional architecture confers resistance to apoptosis in normal and malignant mammary epithelium. *Cancer Cell,* 2002. **2**(3), 205–216.
24. Goel, H. L., et al., beta1 integrins mediate resistance to ionizing radiation in vivo by inhibiting c-Jun amino terminal kinase 1. *J Cell Physiol*, 2013. **228**(7), 1601–1609.
25. Seguin, L., et al., An integrin beta(3)-KRAS-RalB complex drives tumour stemness and resistance to EGFR inhibition. *Nat Cell Biol*, 2014. **16**(5), 457–468.
26. Levental, K. R., et al., Matrix crosslinking forces tumor progression by enhancing integrin signaling. *Cell,* 2009. **139**(5), 891–906.
27. Olive, K. P., et al., Inhibition of Hedgehog signaling enhances delivery of chemotherapy in a mouse model of pancreatic cancer. *Science*, 2009. **324**(5933), 1457–1461.
28. Bergamaschi, A., et al., Extracellular matrix signature identifies breast cancer subgroups with different clinical outcome. *J Pathol*, 2008. **214**(3), 357–367.
29. Naba, A., et al., The matrisome: in silico definition and in vivo characterization by proteomics of normal and tumor extracellular matrices. *Mol Cell Proteomics*, 2012. **11**(4), M111 014647.
30. Pardoll, D. M., The blockade of immune checkpoints in cancer immunotherapy. *Nat Rev Cancer*, 2012. **12**(4), 252–264.
31. Shojaei, F., et al., G-CSF-initiated myeloid cell mobilization and angiogenesis mediate tumor refractoriness to anti-VEGF therapy in mouse models. *Proc Natl Acad Sci U S A*, 2009. **106**(16), 6742–6747.
32. Bayne, L. J., et al., Tumor-derived granulocyte-macrophage colony-stimulating factor regulates myeloid inflammation and T cell immunity in pancreatic cancer. *Cancer Cell,* 2012. **21**(6), 822–835.
33. Motz, G. T., and G. Coukos, The parallel lives of angiogenesis and immunosuppression: cancer and other tales. *Nat Rev Immunol*, 2011. **11**(10), 702–711.

34. Komohara, Y., et al., The coordinated actions of TIM-3 on cancer and myeloid cells in the regulation of tumorigenicity and clinical prognosis in clear cell renal cell carcinomas. *Cancer Immunol Res*, 2015.
35. Finak, G., et al., Stromal gene expression predicts clinical outcome in breast cancer. *Nat Med*, 2008. **14**(5), 518–527.
36. Weizman, N., et al., Macrophages mediate gemcitabine resistance of pancreatic adenocarcinoma by upregulating cytidine deaminase. *Oncogene*, 2014. **33**(29), 3812–3819.
37. Yang, C., et al., Increased drug resistance in breast cancer by tumor-associated macrophages through IL-10/STAT3/bcl-2 signaling pathway. *Med Oncol*, 2015. **32**(2), 352.
38. Wan, S., et al., Tumor-associated macrophages produce interleukin 6 and signal via STAT3 to promote expansion of human hepatocellular carcinoma stem cells. *Gastroenterology*, 2014. **147**(6), 1393–1404.
39. Simons, M., and G. Raposo, Exosomes - vesicular carriers for intercellular communication. *Curr Opin Cell Biol*, 2009. **21**(4), 575–581.
40. Schageman, J., et al., The complete exosome workflow solution: from isolation to characterization of RNA cargo. *Biomed Res Int*, 2013. **2013**, 253957.
41. Peinado, H., et al., Melanoma exosomes educate bone marrow progenitor cells toward a pro-metastatic phenotype through MET. *Nat Med*, 2012. **18**(6), 883.
42. Aung, T., et al., Exosomal evasion of humoral immunotherapy in aggressive B-cell lymphoma modulated by ATP-binding cassette transporter A3. *Proc Natl Acad Sci U S A*, 2011. **108**(37), 15336–15341.
43. Ciravolo, V., et al., Potential role of HER2-overexpressing exosomes in countering trastuzumab-based therapy. *J Cell Physiol*, 2012. **227**(2), 658–667.
44. Luga, V., et al., Exosomes mediate stromal mobilization of autocrine Wnt-PCP signaling in breast cancer cell migration. *Cell*, 2012. **151**(7), 1542–1556.
45. Boelens, M. C., et al., Exosome transfer from stromal to breast cancer cells regulates therapy resistance pathways. *Cell*, 2014. **159**(3), 499–513.
46. Lugini, L., et al., Immune surveillance properties of human NK cell-derived exosomes. *J Immunol*, 2012. **189**(6), 2833–2842.
47. Naslund, T., et al., Dendritic cell derived exosomes need to activate both T and B cells to induce antitumor immunity. *J Immunol*, 2013. **190**(6), 2712–2719.

48. Ozdemir, B. C., et al., Depletion of carcinoma-associated fibroblasts and fibrosis induces immunosuppression and accelerates pancreas cancer with reduced survival. *Cancer Cell,* 2014. **25**(6), 719–734.

49. Rhim, A. D., et al., Stromal elements act to restrain, rather than support, pancreatic ductal adenocarcinoma. *Cancer Cell,* 2014. **25**(6), 735–747.

50. Obenauf, A. C., et al., Therapy-induced tumour secretomes promote resistance and tumour progression. *Nature,* 2015. **520**(7547), 368–372.

51. Chabner, B. A., and T. G. Roberts, Jr., Timeline: chemotherapy and the war on cancer. *Nat Rev Cancer,* 2005. **5**(1), 65–72.

52. Sun, Y., and P. S. Nelson, Molecular pathways: involving microenvironment damage responses in cancer therapy resistance. *Clin Cancer Res,* 2012. **18**(15), 4019–4025.

53. Coppe, J. P., et al., Senescence-associated secretory phenotypes reveal cell-nonautonomous functions of oncogenic RAS and the p53 tumor suppressor. *PLoS Biol,* 2008. **6**(12), 2853–2868.

54. Kuilman, T., and D. S. Peeper, senescence-messaging secretome: SMS-ing cellular stress. *Nat Rev Cancer,* 2009. **9**(2), 81–94.

55. Kuilman, T., et al., Oncogene-induced senescence relayed by an interleukin-dependent inflammatory network. *Cell,* 2008. **133**(6), 1019–1031.

56. Acosta, J. C., et al., Chemokine signaling via the CXCR2 receptor reinforces senescence. *Cell,* 2008. **133**(6), 1006–1018.

57. Sun, Y., et al., Treatment-induced damage to the tumor microenvironment promotes prostate cancer therapy resistance through WNT16B. *Nat Med,* 2012. **18**(9), 1359–1368.

58. Gilbert, L. A., and M. T. Hemann, DNA damage-mediated induction of a chemoresistant niche. *Cell,* 2010. **143**(3), 355–366.

59. Gilbert, L. A., and M. T. Hemann, Chemotherapeutic resistance: surviving stressful situations. *Cancer Res,* 2011. **71**(15), 5062–5066.

60. Roodhart, J. M. L., et al., Mesenchymal stem cells induce resistance to chemotherapy through the release of platinum-induced fatty acids. *Cancer Cell,* 2011. **20**(3), 370–383.

61. Shaked, Y., et al., Rapid chemotherapy-induced acute endothelial progenitor cell mobilization: Implications for antiangiogenic drugs as chemosensitizing agents. *Cancer Cell,* 2008. **14**(3), 263–273.

62. Duan, C. W., et al., Leukemia propagating cells rebuild an evolving niche in response to therapy. *Cancer Cell,* 2014. **25**(6), 778–793.

63. Kim, M., et al., CXCR4 signaling regulates metastasis of chemoresistant melanoma cells by a lymphatic metastatic niche. *Cancer Res*, 2010. **70**(24), 10411–10421.

64. Mannino, M., and A. J. Chalmers, radioresistance of glioma stem cells: intrinsic characteristic or property of the 'microenvironment-stem cell unit'? *Mol Oncol*, 2011. **5**(4), 374–386.

65. Zoni, E., et al., Epithelial plasticity in cancer: unmasking a microRNA network for TGF-beta-, notch-, and Wnt-mediated EMT. *J Oncol*, 2015. **2015**, 198967.

66. Landsberg, J., et al., Melanomas resist T-cell therapy through inflammation-induced reversible dedifferentiation. *Nature*, 2012. **490**(7420), 412–416.

67. Bhat, K. P., et al., Mesenchymal differentiation mediated by NF-kappaB promotes radiation resistance in glioblastoma. *Cancer Cell*, 2013. **24**(3), 331–346.

68. Dijkgraaf, E. M., et al., Chemotherapy alters monocyte differentiation to favor generation of cancer-supporting M2 macrophages in the tumor microenvironment. *Cancer Res*, 2013. **73**(8), 2480–2492.

69. Klug, F., et al., Low-dose irradiation programs macrophage differentiation to an iNOS(+)/M1 phenotype that orchestrates effective T cell immunotherapy. *Cancer Cell*, 2013. **24**(5), 589–602.

70. Tripathi, C., et al., Macrophages are recruited to hypoxic tumor areas and acquire a pro-angiogenic M2-polarized phenotype via hypoxic cancer cell derived cytokines Oncostatin M and Eotaxin. *Oncotarget*, 2014. **5**(14), 5350–5368.

71. Chapman, P. B., et al., Improved survival with vemurafenib in melanoma with BRAF V600E mutation. *N Engl J Med*, 2011. **364**(26), 2507–2516.

72. Young, K., A. Minchom, and J. Larkin, BRIM-1, -2 and -3 trials: improved survival with vemurafenib in metastatic melanoma patients with a BRAF(V600E) mutation. *Future Oncol*, 2012. **8**(5), 499–507.

73. Meric-Bernstam, F., and G. B. Mills, Overcoming implementation challenges of personalized cancer therapy. *Nat Rev Clin Oncol*, 2012. **9**(9), 542–548.

74. Saintigny, P., and J. A. Burger, Recent advances in non-small cell lung cancer biology and clinical management. *Discov Med*, 2012. **13**(71), 287–297.

75. Curigliano, G., New drugs for breast cancer subtypes: targeting driver pathways to overcome resistance. *Cancer Treat Rev*, 2012. **38**(4), 303–310.

76. Van Cutsem, E., et al., Fluorouracil, leucovorin, and irinotecan plus cetuximab treatment and RAS mutations in colorectal cancer. *J Clin Oncol*, 2015. **33**(7), 692–700.
77. Sonkin, D., et al., Tumor suppressors status in cancer cell line Encyclopedia. *Mol Oncol*, 2013. **7**(4), 791–798.
78. Barretina, J., et al., The Cancer Cell Line Encyclopedia enables predictive modelling of anticancer drug sensitivity. *Nature*, 2012. **483**(7391), 603–607.
79. McMillin, D. W., J. M. Negri, and C. S. Mitsiades, The role of tumour-stromal interactions in modifying drug response: challenges and opportunities. *Nat Rev Drug Discovery*, 2013. **12**(3), 217–228.
80. Unger, F. T., I. Witte, and K. A. David, Prediction of individual response to anticancer therapy: historical and future perspectives. *Cell Mol Life Sci*, 2015. **72**(4), 729–757.
81. Roth, M. D., and A. Harui, Human tumor infiltrating lymphocytes cooperatively regulate prostate tumor growth in a humanized mouse model. *J Immunother Cancer*, 2015. **3**, 12.
82. Morton, J. J., et al., XactMice: humanizing mouse bone marrow enables microenvironment reconstitution in a patient-derived xenograft model of head and neck cancer. *Oncogene*, 2016. **35**(3), 290–300.
83. Hovelson, D. H., et al., Development and validation of a scalable next-generation sequencing system for assessing relevant somatic variants in solid tumors. *Neoplasia*, 2015. **17**(4), 385–399.
84. Holohan, C., et al., Cancer drug resistance: an evolving paradigm. *Nat Rev Cancer*, 2013. **13**(10), 714–726.
85. Alspach, E., et al., p38MAPK plays a crucial role in stromal-mediated tumorigenesis. *Cancer Discov*, 2014. **4**(6), 716–729.
86. Laberge, L., et al., mTOR regulates the tumor-promoting senescence-associated secretory phenotype. *Nat Cell Biol*, 2015. **17**, 1049–1061.

Chapter 8

Personalized Immune Therapy

Joost Hegmans, Lysanne Lievense, and Joachim Aerts
Erasmus MC Cancer Institute, Department of Pulmonary Medicine,
Dr. Molewaterplein 50, 3015 GD Rotterdam, The Netherlands
j.aerts@erasmusmc.nl

It has long been assumed that stimulating the patient's immune system to attack tumors is rather meaningless, but this assumption has now changed [1]. Development with therapeutic antibodies, cancer vaccines, and cell-based immunotherapeutic approaches reveals its promise but also the relative infancy of the cancer immunotherapy field. In order to take full advantage of the promise and potential of cancer immunotherapy, research is being focused on the understanding of the patient's immune status for predicting treatment outcome. Therefore, the presence and functional status of various immunological cell types within the tumor microenvironment of the patient is important. This "personalized" knowledge facilitates the identification of the best opportunity for intervention from the plethora of immunotherapeutic approaches currently being developed.

Cancer Genetics and Genomics for Personalized Medicine
Edited by Il-Jin Kim

ISBN 978-981-4669-87-0 (Hardcover), 978-1-315-36476-6 (eBook)
www.panstanford.com

8.1 Immunotherapy

Cancer immunotherapy consists of approaches that enhance the host immune system to generate effective immune responses against cancer. The editors of *Science* have chosen this strategy to combat tumors as the Breakthrough of the Year for 2013 [2]. Although it is still in its infancy, over the last decade we have witnessed that immunotherapeutic approaches are becoming more appealing components of the anticancer armamentarium. In 2010, sipuleucel-T became the first therapeutic vaccine approved by the US FDA for the treatment of metastatic and hormone-refractory prostate cancer [3]. This was followed in 2011 by ipilimumab, a fully human monoclonal antibody that blocks cytotoxic T lymphocyte–associated antigen 4 (CTLA-4), the first agent approved in the EU for the treatment of unresectable or metastatic melanoma [4]. With the increasing understanding of the fundamentals of cellular and molecular tumor immunology, many ways are now being investigated to find out how the immune system can be further augmented to treat cancer. However, to understand the principles of these novel immunotherapeutic approaches, it is important to comprehend the different immune cell components and their function in the tumorigenic process. Some of these immune cells are discussed in the next section, where it becomes clear that the immune system plays a dual role in carcinogenesis. The complex interactions between diverse immune cell types and tumor cells can actively favor tumor rejection as well as tumor progression, depending on the tumor cell characteristics, type, stage, secreted factors, and the types of immune cells that are involved. It emphasizes the importance of the full understanding of the intricacy of the cellular interactions within the tumor microenvironment as well as the effects of tumor-derived factors on surrounding (immunological) cells and tissues. Understanding the local and systemic immune mechanisms will lead to new potential therapeutic targets and better personalized prediction of which patients may benefit from a certain treatment.

8.2 Immune Cell Involvement during Carcinogenesis

Numerous immunological cell types of the host influence cancer incidence, cancer growth, response to therapy, and thereby the prognosis of the disease. Immunohistochemical analysis of cancer biopsies have demonstrated that the tumor microenvironment is a heterogeneous and complex system of tumor cells and stromal cells, including endothelial cells and their precursors, pericytes, smooth-muscle cells, and fibroblasts of various phenotypes, located within the connective tissue or extracellular matrix (e.g., collagen). Leukocyte infiltration is an important characteristic of cancer and the main components of these infiltrates include natural killer (T) cells, mast cells, neutrophils, B- and T-lymphocyte subsets, myeloid-derived suppressor cells, macrophages, and dendritic cells. On the basis of their functions, these cells can be divided into cells with a potentially positive impact on the antitumor response and cells with a detrimental effect (Fig. 8.1). The net effect of the interactions between these various cell types and their secreted products within the environment of an established tumor participates in determining antitumor immunity, angiogenesis, metastasis, overall cancer cell survival, and proliferation. For an update on the function of these immune cells for lung cancer, see the article by Heuvers et al. [5].

8.3 Context-Specific Nature of Immune Cells within Tumors

From the above it is clear that numerous types of tumor-infiltrating immune cells have different effects on tumor progression. On the basis of their functions, immune types can be divided into cells with a potentially positive impact on the antitumor response and cells with detrimental effects, but cell phenotypes can adapt on the changing environment (e.g., the macrophage M1/M2 phenotype polarization). The net effect of the interactions between all these various cell types and their secreted products within the

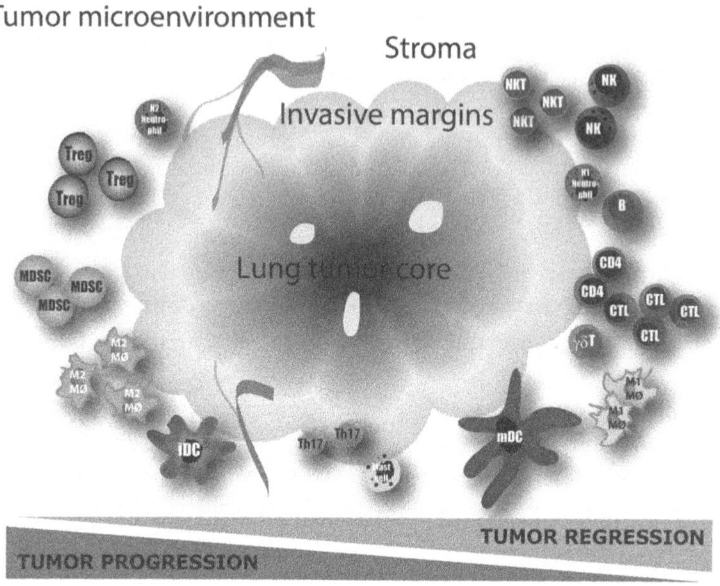

Figure 8.1 The tumor microenvironment is a heterogeneous and complex system of tumor cells and "normal" stromal cells, including endothelial cells and their precursors, pericytes, smooth-muscle cells, and fibroblasts (FBs) of various phenotypes, located within the connective tissue or extra-cellular matrix (e.g., collagen). Leukocyte infiltration is an important characteristic of most cancers and the main components of these infiltrates include natural killer (T) cells (NK/NKT), N1/N2 neutrophils, B- and T-lymphocyte subsets, myeloid derived suppressor cells (MDSC), M1/M2 macrophages (MØ), mast cells, and immature dendritic cells (iDC) / mature dendritic cells (mDC). Based on their functions, these cells can be divided into cells with a potentially positive impact on the antitumor response (right) and cells with a detrimental effect (left). The net effect of the interactions between these various cell types and their secreted products within the environment of an established tumor participates in determining antitumor immunity, angiogenesis, metastasis, overall cancer cell survival, and proliferation. Image adapted from Heuvers et al. [5].

environment of an established tumor participates in determining antitumor immunity, angiogenesis, metastasis, overall cancer cell survival and proliferation. These immune infiltrates are heterogeneous between tumor types and are very diverse from patient to patient. The importance of these tumor-infiltrating immune cells for tumorigenesis, and their secreted chemokines and cytokines, was recently acknowledged by revisiting the hallmarks of cancer as described by Hanahan and Weinberg in 2000 and now include "tumor-promoting inflammation" and "avoiding immune destruction" [6, 7]. Besides the type and density of the cells, also the location in the tumor (in the center or core, or within the invasive margins of the tumor), in adjacent tertiary or secondary lymphoid structures, or their presence in peripheral blood will shape the immune contexture. Histopathological analyses of the location, density, and functional orientation of the different immune cell populations in large annotated collections of human lung tumors have identified a good association of effector T cells (CD3+CD8+), memory T cells (CD3+CD45RO+) [8–12], and Th1 cells (IL-2 and IFN-gamma secreting CD3+ cells) [8] with longer disease-free survival and/or a better overall survival, while Th17 [8, 13] and Tregs [8, 14–16] have a poor association with the prognosis [17, 18]. However, it is important to realize the complex spatiotemporal dynamics in the tumor-immune interactions in time due to perturbations at the gene and protein level of the immune cells and tumor cells within the microenvironment [18]. Thus, the cancer immunoediting concept, as described by Schreiber et al. [19], that comprises editing, equilibrium, and escape, takes place continuously and this plasticity is a major source of heterogeneity between patients and a cause for therapy resistance [20].

8.4 Types of Immunotherapeutic Approaches

Immunotherapy attempts to stimulate or restore the body's natural ability of the immune system to fight cancer. There are various strategies to activate the immune system, and these are classified earlier by Aerts et al. and here into the following categories: biological response modifiers, monoclonal antibodies, peptide or tumor cell vaccines, and cellular immunotherapy. There is no consensus regarding which of the four categories is the optimal approach for malignancies. This is probably highly dependent on the tumor characteristics of each individual patient.

8.4.1 Biological Response Modifiers

Biological response modifiers are compounds which can nonspecifically enhance the immune response, either by directly stimulating the immune system or by the direct induction of tumor cell apoptosis. These compounds can activate the antitumor immune response via the direct stimulation of pro-inflammatory immune cells or via the inhibition of detrimental suppressive immune cells like Tregs or MDSCs. The response modifiers can be divided in compounds that can trigger inflammation, cytokines, or colony-stimulating factors or multimodal effectors.

8.4.2 Monoclonal Antibodies

Monoclonal antibodies bind specifically to one epitope and their application as potential immunotherapeutic agents has received a lot of attention recently. The use of monoclonal antibodies directed against tumor growth related antigens on the tumor cell like epidermal growth factor receptor (EGFR) and human epidermal growth factor receptor 2 (HER2) has been well established in lung cancer patients [21, 22]. In addition to the direct effect of the inhibition of growth factors and/or their receptors, antibodies bound to the tumor cell surface can induce antibody-dependent cell-mediated cytotoxicity (ADCC) [23].

In addition to monoclonal antibodies directed against antigens specifically expressed by tumor cells, antibodies that are directed against tumor products as VEGF (bevacizumab) have been clinically implemented.

The blockade of immune checkpoints using monoclonal antibodies can be considered one of the major breakthroughs in cancer research of the past years. In order to control the immune response and to mitigate collateral tissue damage the immune system is harnessed with a negative feedback system. T cells have the capacity to upregulate co-inhibitory receptors in order to inhibit the immune response and mediate immune tolerance. In chronic infection and cancer, expression of these inhibitory co-receptors is enhanced and associated with an anergic state in T cells [24]. Antibodies that bind to these co-receptors can block inhibitory signals and therefore augment T-cell activation and proliferation. Examples of compounds directed against the immune checkpoints CTLA-4 are ipilimumab and tremelimumab, or against the programmed death protein (PD1) and its ligand PD-L1 as nivolumab. Although these results are without any doubt very promising, the implementation of these immune checkpoint inhibitors is hampered by serious immune-related toxicities (e.g., colitis and hepatitis) and low response rates. Therefore, the development of robust and predictive biomarkers is pivotal for the clinical implementation of monoclonal antibodies against co-inhibitory receptors [25].

8.4.3 Tumor Vaccines

The research regarding cancer vaccines has made great progress since the discovery of human tumor antigens which can be recognized by T-cell receptors [26]. Tumor vaccines are designed to deliver tumor antigens to antigen-presenting cells, which can subsequently induce a tumor specific immune response by the adaptive immune system. These vaccines can consist of various types of antigen sources. An antigen candidate needs to meet certain criteria in order to potentially be able to elicit a specific antitumor immune response. Tumor specificity, frequency, and homogeneous expression in tumor cells, their role as an oncogene, and intrinsic immunogenicity are essential features of antigens which determine the success of vaccines [27]. In lung cancer and mesothelioma, a broad spectrum of approaches using various antigen sources have been undertaken to develop cancer vaccines. These can be divided into (i) proteins and peptides, e.g., GV1001; (2) liposomal complexes, e.g., L-BLP25 or Stimuvax; (3) recombinant viruses and bacterial vectors, e.g., CRS-207; and (iv) cell-based vaccines as GVAX.

8.4.4 Cellular Immunotherapy

Cellular immunotherapy includes the adoptive transfer of autologous or allogeneic activated immune cells. The most prominent success story regarding cellular immunotherapy is sipuleucel-T, a vaccine for prostate cancer that consists of autologous peripheral blood mononuclear cells (PBMCs), including antigen-presenting cells that have been activated ex vivo with a recombinant fusion protein (PA2024, a prostate antigen that is fused to GM-CSF) [3]. After it was demonstrated in a phase 3 clinical trial that sipuleucel-T prolongs survival in metastatic castration-resistant prostate cancer patients, FDA approval followed in 2010. The general goal of adoptive cellular immunotherapy is to induce a tumor-specific immune response via the infusion of, for instance, tumor-antigen-loaded DCs or specifically activated T cells. Cellular immunotherapeutic approaches using various cell types have been evaluated [28].

8.4.4.1 Dendritic cell-based immunotherapy

As described earlier, DCs are the professional antigen-presenting cells of the immune system and they have emerged as the most powerful initiators of immune responses. Because of their capacity to engulf tumor antigens and activate T cells in an antigen-specific manner, the use of DCs as immunotherapeutic agents is very promising. In DC-based immunotherapeutic approaches, DCs are generated ex vivo from monocytes and, after arming with tumor-associated antigens, reinjected into the patient with the intention to restore proper presentation of tumor-associated antigens and T-cell activation. This concept has been researched in NSCLC and has shown promising results regarding the elicited immune response, safety, and tolerability, despite the small sample sizes of the trials [29–31]. In mesothelioma, treatment with autologous tumor-lysate pulsed DCs was shown to be safe and elicited an antitumor immune response in a phase 1 clinical trial [32].

8.4.4.2 Adoptive genetically modified and/or expanded T-cell therapy

Different sources and activation procedures can be used in specifically harnessing the T-cell response and have been clinically evaluated in lung cancer. Lymphokine-activated killer (LAK) cells are autologous IL-2 stimulated lymphocytes and their application has shown clinical responses in several studies with lung cancer patients [33, 34]. Cytokine-induced killer (CIK) cells are generated in vitro by stimulation of peripheral blood lymphocytes with anti-CD3 antibodies, IL-2, IL-1α, and IFN-γ [35]. CIK cells have proven their clinical potential in multiple solid tumors, including lung cancer [36], and these results warrant future clinical trials [37].

8.4.4.3 Adoptive natural killer (T) cells' transfer

Adoptive transfer of allogeneic, in vitro activated and expanded NK cells from haploidentical donors was proven potentially clinically effective in lung cancer [38]. NKT cells are currently exploited for cancer treatment by harnessing these cells with CD1d agonist ligands [39], or by adoptive transfer of NKT cells activated in vitro [40].

8.5 The Future of Personalized Medicine

For a long time, stimulating the patient's immune system to attack tumors has been viewed as a rather meaningless intervention, an assumption that has radically changed since the recent clinical successes of cancer immunotherapy. However, despite the promising results achieved in some patients, the overall response rates are moderate. It has been shown that the immune contexture (i.e., the type, density, and location of tumor-infiltrating immune cells) can predict the clinical outcome of patients affected by multiple types of cancer, but with consistent intra-patient variations [41]. It is therefore critical to identify the unique immunological profile of individual patients as a means to identify the best-suited immunotherapeutic approach for each patient [42]. This personalized way could greatly improve the efficacy of current immunotherapies. In addition to an individualized treatment plan, the future of immunotherapy in lung cancer patients includes multimodality treatment. Surgery, irradiation, or chemotherapy vigorously reduce the tumor mass and the induced tumor cell death results in tumor antigen exposure. These therapeutic effects enhance the efficacy of immunotherapy and warrant combinatorial treatment approaches.

References

1. Pardoll, D., and C. Drake, Immunotherapy earns its spot in the ranks of cancer therapy. *J Exp Med*, 2012. **209**(2), 201–209.
2. Couzin-Frankel, J., Breakthrough of the year 2013. Cancer immunotherapy. *Science*, 2013. **342**(6165), 1432–1433.
3. Kantoff, P. W., et al., Sipuleucel-T immunotherapy for castration-resistant prostate cancer. *N Engl J Med*, 2010. **363**(5), 411–422.
4. Hodi, F. S., et al., Improved survival with ipilimumab in patients with metastatic melanoma. *N Engl J Med*, 2010. **363**(8), 711–723.
5. Heuvers, M. E., et al., Patient-tailored modulation of the immune system may revolutionize future lung cancer treatment. *BMC Cancer*, 2012. **12**, 580.
6. Hanahan, D., and R. A. Weinberg, The hallmarks of cancer. *Cell*, 2000. **100**(1), 57–70.
7. Hanahan, D., and R. A. Weinberg, Hallmarks of cancer: the next generation. *Cell*, 2011. **144**(5), 646–674.
8. Dieu-Nosjean, M. C., et al., Long-term survival for patients with non-small-cell lung cancer with intratumoral lymphoid structures. *J Clin Oncol*, 2008. **26**(27), 4410–4417.
9. Al-Shibli, K. I., et al., Prognostic effect of epithelial and stromal lymphocyte infiltration in non-small cell lung cancer. *Clin Cancer Res*, 2008. **14**(16), 5220–5227.
10. Hiraoka, K., et al., Concurrent infiltration by CD8+ T cells and CD4+ T cells is a favourable prognostic factor in non-small-cell lung carcinoma. *Br J Cancer*, 2006. **94**(2), 275–280.
11. Ito, N., et al., Prognostic significance of T helper 1 and 2 and T cytotoxic 1 and 2 cells in patients with non-small cell lung cancer. *Anticancer Res*, 2005. **25**(3B), 2027–2031.
12. Kawai, O., et al., Predominant infiltration of macrophages and CD8(+) T Cells in cancer nests is a significant predictor of survival in stage IV nonsmall cell lung cancer. *Cancer*, 2008. **113**(6), 1387–1395.
13. Chen, X., et al., Increased IL-17-producing cells correlate with poor survival and lymphangiogenesis in NSCLC patients. *Lung Cancer*, 2010. **69**(3), 348–354.
14. Petersen, R. P., et al., Tumor infiltrating Foxp3+ regulatory T-cells are associated with recurrence in pathologic stage I NSCLC patients. *Cancer*, 2006. **107**(12), 2866–2872.

15. Shimizu, K., et al., Tumor-infiltrating Foxp3+ regulatory T cells are correlated with cyclooxygenase-2 expression and are associated with recurrence in resected non-small cell lung cancer. *J Thorac Oncol*, 2010. **5**(5), 585–590.
16. Tao, H., et al., Prognostic potential of FOXP3 expression in non-small cell lung cancer cells combined with tumor-infiltrating regulatory T cells. *Lung Cancer*, 2012. **75**(1), 95–101.
17. Fridman, W. H., et al., The immune contexture in human tumours: impact on clinical outcome. *Nat Rev Cancer*, 2012. **12**(4), 298–306.
18. Bindea, G., et al., Spatiotemporal dynamics of intratumoral immune cells reveal the immune landscape in human cancer. *Immunity*, 2013. **39**(4), 782–795.
19. Schreiber, R. D., L. J. Old, and M. J. Smyth, Cancer immunoediting: integrating immunity's roles in cancer suppression and promotion. *Science*, 2011. **331**(6024), 1565–1570.
20. Holzel, M., A. Bovier, and T. Tuting, Plasticity of tumour and immune cells: a source of heterogeneity and a cause for therapy resistance? *Nat Rev Cancer*, 2013. **13**(5), 365–376.
21. Pirker, R., EGFR-directed monoclonal antibodies in non-small cell lung cancer. *Target Oncol*, 2013. **8**(1), 47–53.
22. Mazieres, J., et al., Lung cancer that harbors an HER2 mutation: epidemiologic characteristics and therapeutic perspectives. *J Clin Oncol*, 2013. **31**(16), 1997–2003.
23. Arnould, L., et al., Trastuzumab-based treatment of HER2-positive breast cancer: an antibody-dependent cellular cytotoxicity mechanism? *Br J Cancer*, 2006. **94**(2), 259–267.
24. Pardoll, D. M., The blockade of immune checkpoints in cancer immunotherapy. *Nat Rev Cancer*, 2012. **12**(4), 252–264.
25. Lievense, L. A., J. P. Hegmans, and J. G. Aerts, Biomarkers for immune checkpoint inhibitors. *Lancet Oncol*, 2014. **15**(1), e1.
26. van der Bruggen, P., et al., A gene encoding an antigen recognized by cytolytic T lymphocytes on a human melanoma. *Science*, 1991. **254**(5038), 1643–1647.
27. Tartour, E. Z. L., Lung cancer: potential targets for immunotherapy. *Lancet Respir Med*, 2013. **1**, 551–563.
28. Zheng, Y. W., et al., Current adoptive immunotherapy in non-small cell lung cancer and potential influence of therapy outcome. *Cancer Invest*, 2013. **31**(3), 197–205.

29. Perroud, M. W., Jr., et al., Mature autologous dendritic cell vaccines in advanced non-small cell lung cancer: a phase I pilot study. *J Exp Clin Cancer Res*, 2011. **30**, 65.

30. Wang, K., et al., An autologous therapeutic dendritic cell vaccine transfected with total lung carcinoma RNA stimulates cytotoxic T lymphocyte responses against non-small cell lung cancer. *Immunol Invest*, 2009. **38**(7), 665–680.

31. Zhou, Q., et al., A dendritic cell-based tumour vaccine for lung cancer: full-length XAGE-1b protein-pulsed dendritic cells induce specific cytotoxic T lymphocytes in vitro. *Clin Exp Immunol*, 2008. **153**(3), 392–400.

32. Hegmans, J. P., et al., Consolidative dendritic cell-based immunotherapy elicits cytotoxicity against malignant mesothelioma. *Am J Respir Crit Care Med*, 2010. **181**(12), 1383–1390.

33. Kimura, H., and Y. Yamaguchi, Adjuvant immunotherapy with interleukin 2 and lymphokine-activated killer cells after noncurative resection of primary lung cancer. *Lung Cancer*, 1995. **13**(1), 31–44.

34. Kimura, H., and Y. Yamaguchi, A phase III randomized study of interleukin-2 lymphokine-activated killer cell immunotherapy combined with chemotherapy or radiotherapy after curative or noncurative resection of primary lung carcinoma. *Cancer*, 1997. **80**(1), 42–49.

35. Hontscha, C., et al., Clinical trials on CIK cells: first report of the international registry on CIK cells (IRCC). *J Cancer Res Clin Oncol*, 2011. **137**(2), 305–310.

36. Hui, K. M., CIK cells–current status, clinical perspectives and future prospects–the good news. *Expert Opin Biol Ther*, 2012. **12**(6), 659–661.

37. Li, R., et al., Autologous cytokine-induced killer cell immunotherapy in lung cancer: a phase II clinical study. *Cancer Immunol Immunother*, 2012. **61**(11), 2125–2133.

38. Iliopoulou, E. G., et al., A phase I trial of adoptive transfer of allogeneic natural killer cells in patients with advanced non-small cell lung cancer. *Cancer Immunol Immunother*, 2010. **59**(12), 1781–1789.

39. Dhodapkar, M. V., and J. Richter, Harnessing natural killer T (NKT) cells in human myeloma: progress and challenges. *Clin Immunol*, 2011. **140**(2), 160–166.

40. Motohashi, S., and T. Nakayama, Natural killer T cell-mediated immunotherapy for malignant diseases. *Front Biosci (Schol Ed)*, 2009. **1**, 108–116.

41. Hegmans, J. P., and J. G. Aerts, Immunological profiling as a means to invigorate personalized cancer therapy. *Oncoimmunology*, 2013. **2**(8), e25236.

42. Aerts, J. G., and J. P. Hegmans, Tumor-specific cytotoxic T cells are crucial for efficacy of immunomodulatory antibodies in patients with lung cancer. *Cancer Res*, 2013. **73**(8), 2381–2388.

Chapter 9

Hyperthermic Intraperitoneal Chemotherapy (HIPEC) for Peritoneal Malignancies

G. E. Bates,[a,b] **P. Kim,**[a] **C. M. DeRosa,**[a] **L. Petrukhin,**[a] **Y. Bressler,**[a] **and R. N. Taub**[a]

[a]*Columbia University Medical Center, Department of Medicine, Division of Hematology Oncology, 161 Fort Washington Ave., New York, NY, USA*
[b]*Mesothelioma Applied Research Foundation, 1317 King Street, Alexandria, VA 22314, USA*
geb2125@columbia.edu, gbates@curemeso.org

9.1 What Is Hyperthermic Intraperitoneal Chemotherapy (HIPEC)?

Peritoneal malignancies represent a unique group of cancers that are spread over large internal convoluted surfaces in close proximity to vital abdominal organs. Complete surgical removal of these cancers is difficult to accomplish and verify. Additional measures have thereby been resorted to, such as whole abdominal radiation, or the washing of the peritoneal surfaces by chemotherapeutic solutions introduced into the abdominal cavity. Intracavitary chemotherapy

Cancer Genetics and Genomics for Personalized Medicine
Edited by Il-Jin Kim

ISBN 978-981-4669-87-0 (Hardcover), 978-1-315-36476-6 (eBook)
www.panstanford.com

seems especially attractive because it allows higher concentrations of drug to reach the peritoneal surfaces with less toxicity than intravenous chemotherapy. Recently, evidence has accumulated suggesting intracavitary drugs may be even more effective if they are heated upon introduction to the body.

Intraoperative hyperthermic intraperitoneal chemotherapy (HIPEC) is now advocated by numerous experienced mesothelioma specialists as part of the standard of care for malignant peritoneal mesothelioma (MPM). Intraoperative chemotherapy administration is understood to provide better fluid distribution than locally administered peritoneal catheter chemotherapy which may be degraded by postoperative adhesions and the development of preferential intraperitoneal pathways for perfusion fluid (Averbach, 1996).

Interestingly, heat alone has demonstrated cytotoxic capabilities in vitro, independent of its known synergism with certain cytotoxic drugs (Streffer, 1990). Hyperthermia may also allow greater penetration of chemotherapy into the tissues of the peritoneum (Raaphorst, 1989). Intraoperative hyperthermic intraperitoneal chemotherapy (HIPEC) is now advocated by experienced mesothelioma specialists as standard of care for malignant peritoneal mesothelioma (MPM).

There are various protocols for the administration of HIPEC. Generally, heated chemotherapy is instilled as a solution, intra-

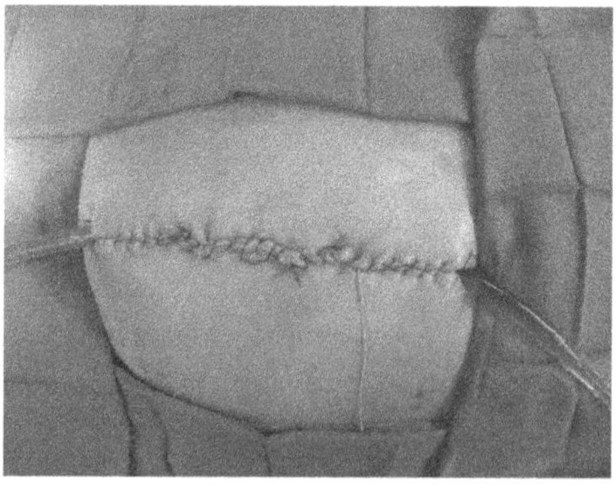

Figure 9.1 HIPEC closed treatment. Photo credit: Michael D. Kluger, Department of Surgery, Columbia University Medical Center.

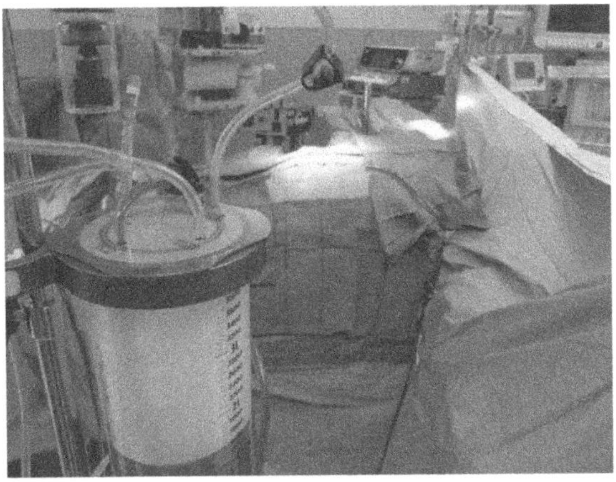

Figure 9.2 HIPEC closed with perfusate. Photo credit: Michael D. Kluger, Department of Surgery, Columbia University Medical Center.

operatively, after tumor debulking. Temperature for administration ranges from 40–42.5°C (varying between institutions but constant during surgery) and are administered for 60–120 minutes (also varying among institutions but constant for each institution). Open, closed, and partially closed techniques have been described with different methods to maintain hyperthermic flow—defined by even fluid distribution, and maximal contact of the instilled drug with the peritoneal surfaces (Dahlke, 2007; Witkamp, 2001). Depending on institutional protocol, after said time, HIPEC perfusate is removed from the peritoneal cavity (Deraco, 2008). These variations in surgical technique opacifies the outcome of HIPEC and makes it a subject rich for research.

9.2 The Role of Heat as a Cytotoxic Agent

The earliest known study of oncological hyperthermic treatment was performed in 1903 by Loeb and Jensen. They found that the growth of malignant sarcomas was inhibited in rats after 30 minutes of exposure at 45°C. Confirmed a few years later by Haaland in 1907, hyperthermia quickly became a subject of interests among scientists.

Table 9.1 Chronology of hyperthermic practices

Author	Whole-body hyperthermia technique	Year
Kapp and Lord	Water bath immersion	1982
Oleson	Magnetic induction coil	1984
Macy	Humidified hyperthermic cabin	1985
Thrall	Radiant heat device	1986
Thrall	Annular-array microwave device	1990

In 1912, Dr. RA Lambert conducted what may have been the first round of experiments comparing normal connective tissue to malignant tumors. In one such study both types of tumors were exposed to temperatures ranging from 42–47°C, Lambert found that mouse sarcoma cells survived 3 hours of continuous exposure at 43°C whereas normal connective tissue survived twice as long (Streffer, 2006).

In 1982, Dr. Kapp and Dr. Lord began the first in vivo experiments investigating hyperthermia as a treatment for cancer on tumor bearing dogs. Over the next two decades, investigators continued exploring the use of heat for the treatment of cancer. Various modalities were used to administer heat (Table 9.1). Also, researchers found the use of localized heat provided an advantage by enhancing local tissue penetration of drug.

9.3 The Role of Heat Shock Proteins in HIPEC Treatment

The growing understanding of cancer genetics has led to the rise of better targeted therapies, improving the quality of life for patients suffering from metastases. This shift in developmental therapies has brought attention to a vibrant aspect of cancer biology: oncogenic proteins that regulate cell proliferation and survival (Goetz, 2003).

It was under hyperthermic conditions that heat shock proteins (HSPs) were discovered; their expression results from exposure to high temperatures in the cellular environment (Aispe, 2004). These proteins, which also respond to other cellular stressors (such as nutrition deprivation), prevent denaturation of intracellular proteins. Commonly referred to as molecular chaperone proteins, HSPs may augment tumorigenesis by allowing neoplastic cells to survive hostile environments. First discovered as a physical and chemical stressor, these highly conserved proteins are upregulated during times of cellular stress and promote various chaperoning properties which serve either a protective function or as a stimulus to the immune system (Goetz, 2003; Aispe, 2004).

There are many chaperone families of HSPs, conventionally categorized by their molecular weight in kilodaltons. HSPs differ in cellular localization and functions; their roles include: controlling regulatory proteins, refolding misfolding proteins, and repairing or degrading proteins damaged by environmental stressors. Protein chaperones have long been associated with tumor cells. HSPs have been understood to have an important protective effect on such cells (Goetz, 2003; Morano, 2007).

9.4 Chemotherapeutic Agents Used for HIPEC

9.4.1 Cisplatin: [cis-diamminedichloroplatinum(II) (CDDP)] (Platinol and Platinol-AQ)

Cisplatin is a platinum based chemotherapeutic drug which works by cross-linking DNA inter-strands, thereby inhibiting DNA duplication, essential for cell division (mitosis). Damaged DNA strands consequentially trigger repair mechanisms which inevitably fail, leading to apoptosis of the cell.

Cisplatin can be administered intravenously (IV) or through intrapleural/intraperitoneal (IP) injection. Since platinum is insoluble in aqueous solution, it is administered in a chelated form (cis-$[Pt(NH_3)_2(Cl)_2]$). Once in the body, one of its chloride ligands is slowly displaced by water (an aqua ligand) in a process termed aquation. The aqua ligand in the resulting $[Pt(NH_3)_2(H_2O)(Cl)]^+$ is itself easily displaced, allowing the platinum atom to bind to nitrogenous base of DNA (guanine is preferred). Subsequent to formation of $Pt(NH_3)_2(Cl)(guanine\text{-}DNA)]^+$, cross linking can occur via displacement of Cisplatin's second chloride ligand, typically by another guanine (Trzaksa, 2005).

Most notable among the changes in DNA are the 1, 2-intrastrand cross-links with purine bases. These include 1, 2-intrastrand d(GpG) adducts which account for nearly 90% of adducts found (from cisplatin) and the less common 1, 2-intrastrand d(ApG) adducts. 1, 3-intrastrand d(GpXpG) adducts occur too but are readily excised by nucleotide excision repair. Other adducts include inter-strand crosslinks and nonfunctional adducts that have been postulated to contribute to cisplatin's activity. Interaction with cellular proteins, particularly HMG domain proteins, has also been advanced as a secondary mechanism of cisplatin's interference with mitosis.

Although cisplatin is frequently designated as an alkylating agent, it has no alkyl group and therefore cannot carry out alkylating reactions. It is correctly classified as alkylating-like agent.

9.4.2 Doxorubicin: (7S,9S)-7-[(2R,4S,5S,6S)-4-amino-5-hydroxy-6-methyloxan-2-yl]oxy-6,9,11-trihydroxy-9-(2-hydroxyacetyl)-4-methoxy-8,10-dihydro-7H-tetracene-5,12-dione)

Doxorubicin, also hydroxyldaunorubicin, is a DNA-interacting drug widely used in chemotherapy. It is an anthracycline antibiotic which is structurally similar to daunomycin, both of which intercalate DNA. Doxorubicin is used in the treatment of a wide range of cancers and is administered by injection. The exact mechanism of action of doxorubicin is complex and still somewhat unclear, though it is understood to interact with DNA by intercalation, thereby inhibiting macromolecular biosynthesis. The planar aromatic chromophore portion of the doxorubicin molecule intercalates between two base pairs of DNA, while six-membered daunosamine sugar sits in the minor groove and interacts with flanking base pairs immediately adjacent to the intercalation site, as evidenced by crystallography.

This, in turn, inhibits the function of enzyme topoisomerase II (which unwinds DNA for transcription). After it has broken the DNA chain for replication, thus preventing the DNA double helix from being resealed and thereby stopping the process of replication. It may have other functions as well, particularly that of peroxidation of mitochondrial lipid membranes.

9.4.3 Carboplatin: (cis-diammine-1, 1-cyclobutane dicarboxylate platinum II, CBDCA, JM8)

Carboplatin is an analogue of cisplatin, and demonstrates lower toxicity properties when compared with cisplatin in clinical trials, as well as higher antitumor activity.

In 1986, DeGregorio et al. investigated IP carboplatin's pharmacokinetics by delivering Carboplatin at 200 mg/m^2 through an indwelling peritoneal catheter. The drug was reconstituted in a 210.45% normal saline warmed to 30–37°C prior to administration to the peritoneal cavity. The saline used was modified to be half-normal in order to avoid conversion of carboplatin to cisplatin in the presence of high chlorine concentrations.

Pharmacokinetic analysis suggest that because the average AUC for total platinum in the plasma was 5.6% where it was 15.6% for the peritoneum at 4 hours and 24 hours, the ratio of the AUCs for peritoneal fluid compared to plasma proved that intraperitoneal administration is a highly effective method of deliverance. The exposure to the peritoneum averaged 18-fold higher at 4 hours and 6-fold higher at 24 hours than plasma exposure. The ratio for total platinum AUC for 24 hours is suspected to be a reflection of carboplatin's longer half-life in the plasma than the peritoneum. About 64% of the administered drug dose was eliminated by the kidneys in the first 24 hours. 21–33 mL/min was of the total platinum constant renal clearance. Research has since reinforced carboplatin's efficacy through IP administration.

9.4.4 Melphalan: 4-[bis(chloroethyl)amino] phenylalanine]

Melphalan is a typical alkylating agent that contains two functional groups: phenylalanine and an N-Lost moiety. The cytotoxicity of melphalan depends on the reactivity of its free amino group. Recent studies have focused on developing the therapeutic potential of melphalan because of its dose-limiting toxicities and limitations towards cells developing resistance.

Melphalan dose in HIPEC has not been standardized internationally. However, some publications propose dozes of melphalan in 1.5% dextrose peritoneal dialysis solution to be administered at 50–70 mg/m^2 of the BSA as a safe dose for 60–90 minutes of the peritoneal perfusion. Bijelic et al. used this same large dose ranging from 50–70 mg/m^2 of melphalan in HIPEC administered for a shorter period of time of 60 minutes maximum. Although a higher dosage was used, Bijelic et al. seemed to have a lower rate of neutropenia due to toxicity (9% of patients).

When melphalan is heated at 40–45°C, melphalan's cytotoxic efficacy is observed to increase. Thus, heat's alterations to a cell's membrane permeability are crucial to drug-resistant cell lines—a reason for melphalan's use in trials for recurrent malignancies. According to Bijelic et al., melphalan has a highly favorable profile with a broad range of efficacy.

9.5 Quality-of-Life Impact of HIPEC Treatment

The physical, psychological, and social burdens of patients being treated for various types of peritoneal cancer are often accompanied by expenses associated with travel, lodging, and medical treatment. This often becomes overwhelming and burdensome to patients and their families. In particular, the financial expenditures of HIPEC are usually large and can be even larger when treatments are postponed or canceled due to adverse effects of disease or treatment (Houts, 1984). Understanding the impact HIPEC has on patients' quality of life (QOL) has become increasingly more important as the treatment is more often used and survival rates increased.

The outcome of QOL in regard to cytoreductive surgery and HIPEC remains disputed. Investigators have reported significant reduction in QOL for the immediate 6 months post-surgery (Seretis, 2014). Others have reported no change (Chia, 2014). It is important for researchers to reach a conclusion about the outcome of this surgery because the quality of one's life after treatment is just as important as a successful procedure. Both should be taken into account while developing a treatment plan.

The effect of QOL is particularly important for women of childbearing years as the number of women with peritoneal malignancies is steadily rising, bringing down the median age of diagnosis of premenopausal women. Very few (a total of 6) studies have been conducted to evaluate fertility after HIPEC. Needless to say, effects on fertility greatly impact the QOL of women of childbearing age and should be discussed prior to their receiving HIPEC treatment. Clinicians should balance the reproductive desires of their female patients with known prognostic factors of the cancer being treated.

9.6 The Future of HIPEC: The Need for Personalization

The future of HIPEC will depend largely on phase 1 randomized and standardized clinical trial to show that (or if) it is the most effective treatment for peritoneal malignancies. There is much variation of technique for HIPEC among medical centers. Details of such operations (including: open or closed perfusion, temperature range, and time intervals) should be investigated, clarified, and perhaps standardized to maximize the effect of this treatment. Additionally, the field of personalized medicine has allowed addition of various modes of novel treatment for these difficult tumors. The personalization of HIPEC will allow clinicians to target tumor cells with more specific, and perhaps more appropriate, chemotherapeutic agents.

References

Averbach AM, Sugarbaker PH. Methodologic considerations in treatment using intraperitoneal chemotherapy. *Cancer Treat Res.* 1996;82:289–309.

Raaphorst GP. Thermal radiosenzitation in vitro. In: Urano M, Douple E, eds. *Biology of Thermal Potentiation of Radiotherapy*. Utrecht, Tokyo: VSP; 1989:17–52.

Dahlke MH, Schlitt HJ, Piso P. Continuous peritoneal perfusion: techniques, methods and applications. *Cancer Treat Res.* 2007;134:265–273.

Witkamp AJ, de Bree E, Van Goethem R, Zoetmulder FA. Rationale and techniques of intra-operative hyperthermic intraperitoneal chemotherapy. *Cancer Treat Rev.* 2001;27(6):365–374.

Ceelen WP, Flessner MF. Intraperitoneal therapy for peritoneal tumors: biophysics and clinical evidence. *Nat Rev Clin Oncol.* 2010;7(2):108–115.

Deraco M, Bartlett D, Kusamura S, Baratti D. Consensus statement on peritoneal mesothelioma. *J Surg Oncol.* 2008;98(4):268–272.

Streffer C. Biological basis of thermotherapy (with special reference to oncology). In: Gautherie M, ed. *Biological Basis of Oncologic Thermotherapy.* Berlin, Heidelberg, New York: Springer; 1990:1–72.

Trzaksa S. Cisplatin. *C&EN News* 83(25). June 20, 2005. http://pubs.acs.org/cen/coverstory/83/8325/8325cisplatin.html.

Streffer C. Molecular and cellular mechanisms of hyperthermia. In: Seegenschmiedt M, Fessenden P, Vernon, CC, eds. *Thermo-radiotherapy and Thermo-chemotherapy.* New York: Springer, 2006:47–74.

Morano. 2007. Annals of the New York Academy of Sciences.

Aispe et al. 2004. The FASEB Journal.

Goetz et al. 2003. Annals of Oncology.

Chapter 10

Personalized Medicine in Hereditary Cancer Syndromes

Rishi Agarwal[a] **and Vivek Subbiah**[b]
[a]*The Medical Center at Bowling Green, 250 Park St, Bowling Green, KY 42101, USA*
[b]*Department of Investigational Cancer Therapeutics, Division of Cancer Medicine, The University of Texas MD Anderson Cancer Center Houston, 1515 Holcombe Blvd, Houston, TX 77030, USA*
drrishiagarwal@gmail.com, subbiah@mdanderson.org

10.1 Introduction

The exponential increase in cancer genetics research in the last two decades has unraveled genetic mutations which could be inherited. The presence of these aberrations in the germline predispose an affected individual to benign or malignant tumors. These genetic aberrations often manifest early in life and typically affect children, adolescents, and young adults (AYA) in the prime of their lives. Affected individuals often have more than one malignancy. More than fifty such hereditary cancer predisposing syndromes have been identified so far. Better understanding of cancer genetics has also helped deciphering the molecular aberrations and pathways leading to cancer in hereditary cancer syndromes. In the era of targeted

Cancer Genetics and Genomics for Personalized Medicine
Edited by Il-Jin Kim

ISBN 978-981-4669-87-0 (Hardcover), 978-1-315-36476-6 (eBook)
www.panstanford.com

therapy, when several novel agents have now been approved for various malignancies, a door has opened for patients suffering from hereditary cancer syndrome. In spite of better understanding of molecular pathways in hereditary cancer syndromes, the development of targeted therapies has lagged behind in this group of patients. The rarity of these syndromes and inability to build clinical trials in rare subsets of patients and paucity of participation in clinical trials could be one possible reason behind the lack of targeted therapy development in these patients.

Targeted therapy options available for the hereditary cancer syndromes can be broadly divided four categories based the properties of the agents:

1. Agents targeting the PI3K/AKT/mTOR pathway
2. Agents targeting the RAS/RAF/MEK/ERK pathway
3. Anti-angiogenics
4. Modulators of DNA repair mechanism

The hereditary cancer syndromes can also be broadly divided into similar categories depending on the molecular defect:

1. Defect in PI3K/AKT/mTOR pathway: Cowden syndrome, Proteus syndrome, Tuberous sclerosis complex, neurofibromatosis 1 and 2
2. Defect in the RAS/RAF/MEK/ERK pathway: RASopathies, neurofibromatosis 1 and 2
3. Defect in angiogenesis: von Hippel–Lindau disease
4. Defects DNA repair mechanism: hereditary breast and ovarian cancer syndrome, Lynch syndrome, Li–Fraumeni syndrome
5. Defect in growth factor regulation: Gorlin syndrome
6. Others: familial adenomatous polyposis

In this chapter we will review some of more common hereditary cancer syndromes, potential for targeted therapy options from drugs already in development for other diseases, and the status of personalized targeted therapy for these patients.

10.2 Neurofibromatosis Type 1

Neurofibromatosis type 1 (von Recklinghausen disease) is inherited in an autosomal dominant fashion and affects approximately 1:2500 to 1:3000 individuals. The overall risk of malignancies in affected individuals is 5–15%. Tumors commonly associated with neurofibromatosis type 1 include malignant peripheral nerve sheath tumor (MPNST), central nervous system tumors (optic pathway glioma, astrocytoma, and brain stem glioma), soft tissue sarcoma, intestinal gastrointestinal stromal tumors, and rhabdomyosarcoma [1, 2].

NF-1 gene, mapped to chromosome 17q11.2, is mutated in neurofibromatosis type 1. *NF-1* gene encodes a protein neurofibromin which downregulates the activity of P21-RAS, which affects various signaling pathways involved in tumorigenesis, including the mammalian target of rapamycin (mTOR), mitogen-activated protein (MAP) kinase, and stem cell factor/c-Kit signaling pathways (Fig. 10.1) [1, 2]. Lack of neurofibromin activity in NF-1 patients leads to uninhibited activity of the RAS signaling pathways leading to malignancies. The upregulation of the RAS pathway seen here suggests that targeting the RAS signaling pathway (RAF/MAPK/ERK and phosphoinositide-3-kinase [PI3K]/AKT axes) may be a therapeutic option (Fig. 10.1) [1, 2].

The farnesyltransferase inhibitor tipifarnib was one of the first such agents to be tested in patients with neurofibromatosis type 1. Farnesyltransferase (FTase) and geranylgeranyltransferase (GGTase) are prenylttransferases and play a crucial role in post-translational modification of RAS proteins [2]. Tipifarnib, an FTase inhibitor showed promise in preclinical models but no significant objective response was noted in phase 1 and 2 trials [2, 3]. Combination of FTase inhibitors and GGTase inhibitors were found to highly toxic in mouse models and are currently not in development [1].

BRAF inhibitors, MEK inhibitors and mTOR inhibitors have been investigated in NF-1 [1, 2]. Sorafenib (Nexavar) reduced tumor volumes in mouse model of NF-1, but phase 1 trial did not suggest any clinical benefit in treating inoperable plexiform neurofibromas

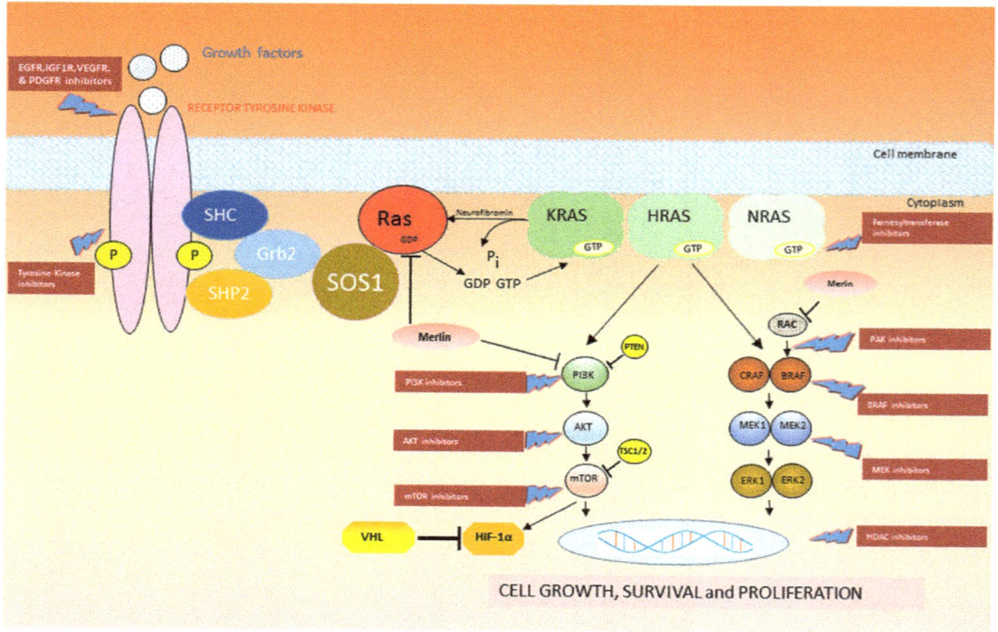

Figure 10.1 PIK3CA/AKT/mTOR and RAS/RAF/MEK/ERK pathway [1, 33, 46]. Aberration involving one or more proteins involved in these pathways are responsible for several hereditary cancer syndromes. Targeted therapy options are highlighted in red.

in NF-1 patients [4]. Sirolimus, which is an mTOR inhibitor, failed to show any significant tumor shrinkage in neurofibromatosis type 1-associated plexiform neurofibroma but a recent consortium study showed that sirolimus prolongs time to progression by almost 4 months in patients with NF1-associated progressive plexiform neurofibroma. However, similar benefit was not seen in patients with nonprogressive plexiform neurofibromas [5]. Several other trials are investigating the role of everolimus for various manifestations of NF1 (NCI trial #NCT01365468; NCI trial #NCT01158651). MEK inhibitors may have a role either alone or in combination with mTOR inhibitors and are currently under investigation (NCI trial #NCT02096471) [1, 2].

c-Kit inhibitors also showed preclinical benefits and subsequently imatinib mesylate (Gleevec) was tested in a phase 2 trial in which 17% (6/36) objective response rate was reported [6]. Other agents targeting the c-Kit mechanisms are under investigation (nilotinib (Tasigna): NCI trial #NCT01275586; sunitinib (Sutent): NCI trial #NCT01402817).

Epidermal growth factor receptor (EGFR) and pro-angiogenic factors are also overexpressed in NF1 affected cells and drugs targeting these mechanisms such EGFR inhibitor (erlotinib) and VEGF inhibitors (Ranibizumab and bevacizumab) are under investigation for NF1 patients (NCI trial #NCT00901849, NCI trial #NCT00657202; NCI trial #NCT01661283) [1, 2].

10.3 Neurofibromatosis Type 2

Neurofibromatosis type 2 (NF-2) is rarer in incidence (1 in 25,000) compared to NF1 [7]. It is caused by a loss of function mutation in the *NF2* gene on chromosome *22q12*, which encodes a tumor suppressor protein called merlin [1, 7, 8]. Clinical manifestations of NF-2 involves the nervous system(vestibular schwannoma, usually bilateral), cranial nerve schwannoma, intracranial meningioma, spinal tumors and peripheral neuropathy, eyes (cataracts, epiretinal membranes, optic nerve meningiomas, and retinal hamartomas), and skin,(cutaneous and subcutaneous tumors as well as skin plaques) [7].

Surgery is the mainstay for management but systemic therapy becomes relevant in patients who are poor surgical candidates. Merlin is involved in multiple pathways involved in cancer development such as RAC, PI3K/AKT/mTOR, RAF/MAPK, and it also controls the surface availability of membrane receptors such as VEGF, Notch, EGFR, and other members of the ErbB family [8]. Several novel agents targeting these pathways may have potential role, but no such therapy has been approved for NF2 patients so far (Fig. 10.1) [1, 8, 9].

Bevacizumab, a VEGF inhibitor, was shown to produce tumor regression as well as improvement in hearing in multiple patients with vestibular schwannoma [9, 10]. Erlotinib, an EGFR tyrosine kinase inhibitor, improved hearing and produced radiologic response in one patient, but similar radiologic response was not seen in a larger trial of 11 patients in which 4 of 11 patients showed stable disease [11]. In a phase 2 trial, dual EGFR/HER2 inhibitor, lapatinib led to volumetric response in 4 of 17 patients and hearing improvement in 4 of 7 patients with vestibular schwannoma associated with neurofibromatosis type 2 [12].

Other agents with possible role in NF-2 patients are in various stages of investigation and include the mTOR inhibitors, HER2 inhibitor trastuzumab, and PAK inhibitors [1, 8].

10.4 Gorlin Syndrome

Germline mutations in the human homolog of the patched (*PTCH*) gene, mapped to chromosome *9q22.3* can lead to Gorlin or nevoid basal cell carcinoma syndrome, which is inherited in autosomal dominant fashion and has an incidence of 1:57,000 to 1:256,000 individuals. Common manifestations included development defects, early-onset nevoid basal cell carcinoma and multiple medulloblastomas (multiple lesions) [1, 13].

PTCH gene product is a transmembrane protein receptor which binds to the smoothened transmembrane protein (SMO) and negatively regulates the sonic hedgehog (SHH) signaling pathway. Mutations in the *PTCH* gene in patients with Gorlin syndrome results in decreased synthesis of PTCH protein leading to aberrant

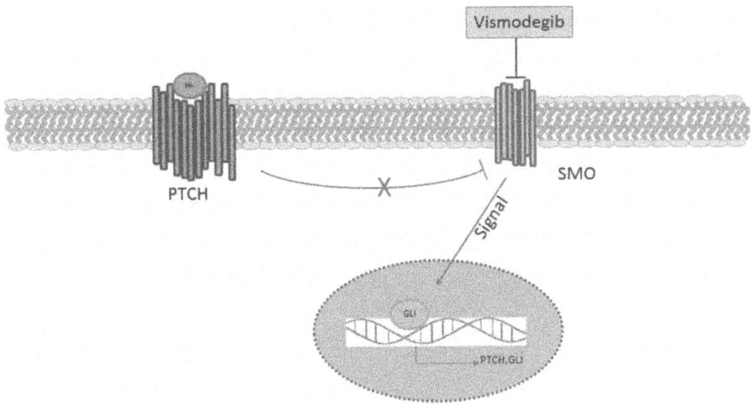

Figure 10.2 Mechanism of sonic hedgehog (SHH) inhibitor (Vismodegib) as targeted therapy for Gorlin syndrome [1, 15]. Patched protein (PTCH) inhibits smoothened (SMO) receptor. Mutations in the PTCH gene seen in Gorlin syndrome leads to decreased synthesis of PTCH protein, causing aberrant activation of the SHH pathway. Vismodegib and other SMO inhibitors block the Hedgehog pathway by inhibiting SMO.

activation of the SHH pathway, which causes abnormal proliferation of basal cells of skin (Fig. 10.2) [1, 13].

Vismodegib is an oral inhibitor of the SHH pathway and has FDA approval for the treatment of advanced basal cell carcinoma (Fig. 10.2) [1, 14]. In an interim analysis of a randomized, placebo-controlled trial in patients with Gorlin syndrome, vismodegib (GDC-0449) lead to a significant decrease in disease burden [15]. Other oral hedgehog pathway antagonists, BMS-833923 (XL139) and sonedegib (LDE225), have shown clinical activity in patients with Gorlin syndrome [1, 15]. Multiple other hedgehog pathway inhibitors are under clinical investigation, but their role in treating Gorlin syndrome remains unclear [15].

10.5 Hereditary Breast and Ovarian Cancer Syndrome

As the name suggests, patients with hereditary breast and ovarian cancer syndrome (HBOCS) are at increased risk of developing breast

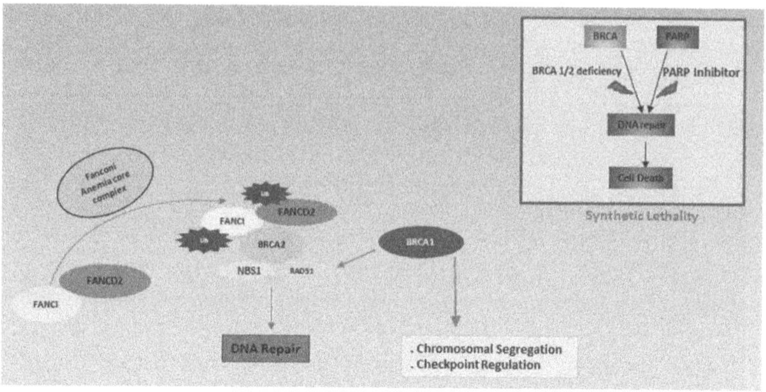

Figure 10.3 The Fanconi anemia/BRCA DNA repair pathway [17, 33]. In cells with BRCA 1/2 deficiency, PARP inhibitors induce cell death through "synthetic lethality" (box).

or ovarian cancer along with other cancers [16, 17]. In addition to breast and ovarian cancers, patients with a *BRCA1* mutation are also at increased risk of developing cervical, uterine, pancreatic, gastric, and prostate cancer, whereas patients with a *BRCA2* mutation are at increased risk of developing melanoma, gallbladder, bile duct, prostate, pancreatic, and stomach cancers [16, 17].

Mutations in *BRCA1* (*17q21*) and *BRCA2* (*13q12.3*) have been linked to this syndrome [16, 17]. *BRCA1* and *BRCA2* functions as tumor suppressor genes responsible for maintaining genomic integrity mostly by three important mechanisms: regulating DNA double-strand break repair with homologous recombination, controlling cell cycle checkpoint responses and segregating chromosomes (Fig. 10.3) [17].

Polyadenosine diphosphate [ADP]-ribose polymerase (PARP) inhibitors have been extensively studied in HBOCS. PARP inhibition impairs base excision repair and leads to DNA breaks, which are repaired by BRCA1, BRCA 2, and PALB2 (for "partner and localizer of BRCA2") mediated homologous recombination in patients with intact BRCA1 and BRCA2 [18, 19]. In *BRCA1- or BRCA2*-deficient patients, BRCA-mediated repair is impaired and addition of PARP inhibitors leads to accumulation of DNA breaks, resulting in loss of cell viability (Fig. 10.3) [17, 18].

Several PARP inhibitors are currently in various phases of clinical development. Recently, olaparib a PARP inhibitor has been FDA approved for advanced ovarian cancer patients with BRCA mutations based on a recently published phase 2 study in which advance cancer patients (n = 298) with germline BRCA1/2 mutations were treated with 400 mg twice a day olaparib [20]. Overall tumor response rate for all 298 patients was 26.2%, which included a tumor response rate of 31% in ovarian cancers, 12% in breast cancers, and 50% in prostate cancers. Similar results in BRCA1/2 mutated ovarian cancer were also reported by Audeh et al. (33% overall response rate) and Fong et al. (46% overall clinical benefit) [17].

PARP inhibitors also increase sensitivity to conventional chemotherapy by preventing DNA repair, suggesting a role for combination regimen. In a randomized phase 2 trial of high grade, recurrent ovarian cancer patients, BRCA-mutated patients had a significantly improved progression-free survival when receiving olaparib-chemotherapy combination vs chemotherapy alone (HR 0.21 [95% CI 0.08–0.55]; p = 0.0015) [21]. Olaparib has also shown activity in combination with Cediranib (tyrosine kinase inhibitor of VEGF receptors 1, 2, and 3) [22]. Another PARP inhibitor veliparib plus temozolomide has also shown clinical benefit of 45% in 20 *BRCA1-or BRCA2*-mutated metastatic breast cancer patients [23]. A randomized phase 2 trial is ongoing, which will compare the combination of veliparib, and temozolomide with the combination of carboplatin plus paclitaxel in *BRCA1*- or *BRCA2*-mutated breast cancer patients (NCI trial #NCT01506609). Several other clinical trials are investigating PARP inhibitors either alone or in combination setting for BRCA mutated cancers (NCI trial #s NCT02163694, NCT01905592, NCT01989546, NCT02401347, NCT02326844, NCT01945775, NCT02396433, NCT01472783, NCT01844986, NCT02034916, NCT02286687, NCT02000622, NCT01506609, NCT02042378, NCT02262273, NCT01585805, NCT01482715, NCT01434316, NCT02354586, NCT02338622, NCT00989651, NCT01017640).

10.6 Lynch Syndrome

Lynch syndrome, or hereditary non-polyposis colorectal cancer (HNPCC) syndrome, accounts for 1–4% of all cases of colon cancer [24]. Along with colon cancer, endometrial, ovarian, upper urologic tract, gastric, small bowel, and biliary/pancreatic cancer are also seen in patients with HNPCC [24].

HNPCC is inherited in autosomal dominant pattern. Germline mutation in 1 of the 4 DNA mismatch repair genes: *MSH2* on chromosome *2p16*, *MLH1* on chromosome *3p21*, *MSH6* on chromosome *2p16* or *PMS2* on chromosome *7p22*, is responsible for this syndrome. Mutations in *MSH2* have been linked to higher incidence of extracolonic tumors. Germline mutations in *EPCAM*, which lies upstream of *MSH2* have also been linked to Lynch syndrome [24]. Mutation in the DNA mismatch repair genes leads to increased mutation rates resulting in increased microsatellite instability. Testing of microsatellite instability by polymerase chain reactions is the diagnostic methods used to identify HNPCC patient.

Nonconventional chemotherapeutic agents such as nonsteroidal anti-inflammatory drugs (NSAIDs) and resistant starch have been studied as potential treatments in patients with Lynch syndrome [17].

Cyclooxygenase-2 (COX-2) is overexpressed in a variety of malignancies, and COX-2 inhibitors have shown to have a protective role in HNPCC patients [25]. A phase 1 trial of COX-2 inhibitors in Lynch syndrome patients is in process (NCI trial #NCT00001693). Role of aspirin (a nonselective COX1/2 inhibitor) was evaluated in the adenoma/carcinoma prevention program (CAPP2) study. At a follow-up of 56 months, HNPCC patients who received aspirin for at least 2 years showed decreases cancer incidence [26].

10.7 Familial Adenomatous Polyposis

Familial adenomatous polyposis (FAP) is an autosomal dominant hereditary colon cancer syndrome which accounts for ≤1% of all patients with colorectal cancer. Along with numerous colorectal adenomatous polyps and early-onset colorectal carcinoma which is characteristic of FAP, there are also extracolonic manifestations such as polyps in other parts of gastrointestinal tract, desmoid tumors, sebaceous or epidermoid cysts, lipoma, osteoma, fibroma, supernumerary teeth, juvenile nasopharyngeal angiofibroma, congenital hypertrophy of the retinal pigment epithelium, and thyroid carcinoma. Gardner syndrome and Turcot syndrome are variants of FAP-associated with multiple extraintestinal manifestations and medulloblastoma, respectively [24].

The underlying mutation associated with FAP and its variants occurs in the adenomatous polyposis coli (*APC*) gene, which is a tumor-suppressor gene, located on chromosome *5q21-q22*. Loss of function mutation in APC gene impairs Wnt signaling-mediated B-catenin degradation, resulting in B-catenin accumulation, which leads to increased cell proliferation [24, 27]. Mutations in *APC* also deleteriously affect cell migration, chromosome stability, and DNA repair [27].

Similar to Lynch syndrome, targeted therapy development in FAP is mostly limited to NSAIDs, which exert antineoplastic effect through inhibition of Wnt-signaling pathway and by preventing accumulation of beta-catenin. Sulindac is one of the first NSAIDSs to show clinical activity in patients with FAP. In one of the initial randomized studies, treatment with sulindac led to significant decrease in number and size of patients with FAP [28]. Celecoxib is another NSAID with reported clinical activity in FAP patients. A randomized controlled trial patients with FAP who were treated with 400 mg BID celecoxib showed 28% reduction in the incidence of colorectal polyps compared to 4.5% reduction in patients treated with a placebo [29]. Aspirin, rectal sulindac, rectal indomethacin, sulindac sulfone, and rofecoxib also play preventative roles in FAP patients [17]. Tiracoxib, however, a selective COX-2 inhibitor, did not show any significant benefit [30].

The possible role of Notch inhibitors have been suggested by preclinical studies where Notch signaling was found to play a role in preventing differentiation of neoplastic intestinal cells in mice with *APC* mutation [31].

Other agents currently under clinical investigation for FAP patients are metformin (NCI trial #NCT01725490), curcumin (NCI trial #NCT00927485), eicosapentaenoic acid (omega-3 fatty acid; NCI trial #NCT00510692), eflornithine (inhibitor of ornithine decarboxylase) + sulindac (NCI trial #NCT01483144), sulindac + erlotinib (EGFR tyrosine kinase inhibitor; NCI trial #NCT01187901), and celecoxib with or without eflornithine (NCI trial #NCT00033371).

10.8 Fanconi Anemia

Fanconi anemia (FA) is a rare autosomal recessive syndrome with an incidence of 1:200,000 to 1:400,000 live births and is characterized by aplastic anemia, congenital abnormalities, and increased risk of leukemia (6–16%) and nonhematologic cancers (5–76%) [32]. Human papilloma virus (HPV)-associated squamous cell carcinomas of the head, neck, cervix, vulva, and skin are more common in patients with Fanconi anemia than in the general population. Other nonhematologic cancers identified include breast, brain, kidney, liver, and pancreatic cancer [32].

Fanconi anemia is a chromosomal fragility disorder caused by germline mutations in genes essential for repair of DNA interstrand cross-links during replication. The FA DNA repair pathway consists of at least 15 FA gene products. The FA core complex formed by eight FA proteins activates the monoubiquitination of FANCD2 and FANCI. The ubiquitinated FANCD2 interacts with several other proteins and coordinates DNA repair activities (Fig. 10.3) [32, 33]. Patients with FA have defects in FA genes, which impairs the repair activities.

Because of the failure of DNA repair mechanism in FA patients, agents targeting the DNA repair pathway such as PARP inhibitors may have a role in FA [33]. Except for one trial (NCI # NCT02286687) with BMN 673 (talazoparib tosylate) that has an arm that allows patients with amplification of EMSY or Fanconi anemia genes (NCI # NCT02286687), no other trials are investigating any targeted therapy in nonhematologic cancer in patients with Fanconi anemia. Cells with defects in Fanconi anemia pathways are more sensitive to damage by chemotherapy agents and radiotherapy. This preclinical hypothesis was substantiated by a report of a pancreatic cancer patient with *PALB2* gene mutation who showed remarkable and prolonged clinical response when treated with DNA-damaging agents mytomycin C and cisplatin [34].

10.9 Inherited Medullary Thyroid Cancer

Inherited medullary thyroid cancer (MTC) is associated with multiple endocrine neoplasia (MEN) type 2a, 2b, and familial MTC. MEN 2a is associated with MTC, pheochromocytoma, and primary parathyroid hyperplasia. MEN2b presents with MTC, pheochromocytoma, marfanoid habitus, mucosal neuromas, and intestinal ganglioneuromatosis. Familial MTC is a variant of MEN that is associated with MTC but lacks other features of MEN syndromes. The underlying genetic defect in the MEN syndrome and familial MTC involves gain of function missense mutations in the *RET* (rearranged during transfection) proto-oncogene on chromosome 10. These mutations are inherited in an autosomal dominant fashion [33, 35].

The *RET* oncogene encodes a transmembrane tyrosine kinase receptor with an extracellular domain, a transmembrane domain, and an intracellular tyrosine kinase region [35]. The gain of function mutation seen in inherited MTC can affect both extracellular and intracellular regions. Mutation in the extracellular cysteine-rich domain leads to ligand independent dimerization of the receptors and downstream activation of the intracellular pathway while mutations in the intracellular tyrosine kinase region leads to constitutive activation of intracellular pathways [33, 35].

Role of targeted therapy has been explored in advanced, unresectable metastatic medullary thyroid cancer. RET kinase inhibitors is the major group of targeted therapy with clinical activity in MTC. Vandetanib and cabozantinib are RET kinase inhibitors which are FDA approved for MTC. Several other multikinase inhibitors such as sorafenib, sunitinib, imatinib, and motesanib can also inhibit RET kinase.

Vandetanib is a once-daily oral tyrosine kinase inhibitor of VEGFR, RET, and EGFR with a recommended dose of 300 mg once daily. The activity of vandetanib in hereditary MTC was shown in a randomized controlled trial in which 331 advanced MTC patients were randomly assigned to receive vandetanib or placebo. Thirty-three patients in the whole group had hereditary MTC and 28 of these received vandetanib. Thirteen of 28 (46%) who received vandetanib showed objective response [36].

Cabozantinib is an oral, small-molecule tyrosine kinase inhibitor that targets VEGFR1, VEGRF2, c-MET, and RET, which is approved by US FDA for advanced metastatic MTC [37]. In the pivotal EXAM trial, 330 patients (46% positive for RET mutations, 5.5% hereditary MTC) were randomized to receive cabozantinib (140 mg per day) or placebo. The median progression-free survival duration was 11.2 months in the treatment group and 4.0 months in the placebo group [37].

Other tyrosine kinase inhibitors which mostly worked through antiangiogenic properties have also shown activity against MTC such as lenvatinib (inhibitor of VEGFR1-3, FGFR1-4, RET, KIT, and PDGFRβ), pazopanib (inhibitor of VEGFR 1-3, PDGFR and c-Kit), and axitinib (inhibitor of VEGF 1-3), but their role in MTC is not clear [33].

10.10 Tuberous Sclerosis Complex (TSC)

TSC is a rare, autosomal dominant, hereditary neurocutaneous disorder affecting the skin and the nervous system and is characterized by widespread hamartomas in multiple organ systems, including the brain, heart, skin, eyes, kidneys, lung, and liver. Patients with TSC are also at increased risk of developing malignancies, including primarily renal cell cancer, angiomyolipoma, lymphangioleiomyomatosis, glioma, and soft tissue sarcoma [38, 39].

Causative mutation in TSC could be on one of following two genes: *TSC1* on chromosome *9q34*, which encodes hamartin protein, or *TSC2* on chromosome *16p13.3*, which encodes protein tuberin. These mutations occur with similar frequency in patients will familial TSC, while *TSC2* mutations are more frequent in sporadic cases [38]. Among the cancer pathways affected by the TSC proteins, mTOR pathway is one of most clinically relevant. mTOR pathway is stimulated by RHEB (Ras homologue enriched in the brain), which itself is downregulated by the tuberin–hamartin complex. In patients with TSC, the inhibition of RHEB by tuberin–hamartin complex is absent, causing constitutive activation of mTOR and its downstream signaling pathways [33, 40]. mTOR inhibitors have been widely studied and are also being utilized for TSC patients.

After mTOR upregulation was found to be associated with TSC, the benefits of mTOR inhibitors to treat TSC were explored. Sirolimus and everolimus (mTOR inhibitor) have been extensively studied and clinically significant activity has been seen for both these agents in phase 2 and phase 3 trials [40]. EXIST-I and EXIST-II trials are the two randomized phase 3 trials which established the role of everolimus in TSC and led to everolimus approval for TSC in patients diagnosed with unresectable subependymal giant cell astrocytoma and TSC who have renal angiomyolipoma that does not require immediate surgery [41, 42].

10.11 RASopathies

RASopathies are a group of syndromes which are caused by genetic mutations in the RAS/RAF/MAPK pathway. This group includes Noonan syndrome, Costello syndrome, cardiofaciocutaneous syndrome, and LEOPARD (**L**entigines, **E**KG abnormalities, **O**cular hypertelorism, **P**ulmonary valve stenosis, **A**bnormal genitalia, **R**etardation of growth, and **D**eafness) syndrome with distinct genetic mutations. These are inherited in an autosomal dominant fashion. Clinically these syndromes have overlapping manifestations, the most common of which are developmental delay, abnormal facial features, skeletal and ectodermal anomalies, congenital heart defects, and increased risk for cancers [43]. These syndromes are also called neurocardiofaciocutaneous syndrome family (Fig. 10.1) [33, 44]. Neuroblastoma, leukemia, rhabdomyosarcoma, and glioma are common cancers that occur in these patients.

The underlying genetic mutations affect different sections of the RAS/RAF/MAPK pathway in each syndrome. Noonan syndrome is associated with mutations in the *PTPN11*, *KRAS*, *SOS*, and *RAF1* genes [44]. LEOPARD syndrome is associated with mutations in *PTPN11* and *RAF1*. Patients with Costello syndrome have mutations in the *HRAS* gene. Patients with cardiofaciocutaneous syndrome have mutations in *BRAF*, *MAP2K1* and *MAP2K2*, and *KRAS* [44].

Developing a targeted therapy for the patients with RAS mutations has been a challenge so far. However, possible roles can be suggested for several agents which act as downstream of RAS, such as farnesyltransferase inhibitors, RAS antagonists, RAF kinase inhibitors, BRAF inhibitors, and MEK inhibitors (Fig. 10.1) [33]. No specific targeted anticancer therapy has been reported so far to have a significant clinical benefit in any of these syndromes, but there are some reports of preclinical activity of MEK inhibitors in Noonan syndrome and cardiofaciocutaneous syndrome [33].

10.12 Von Hippel–Lindau Disease

Von Hippel–Lindau disease (VHL) is an autosomal dominant hereditary cancer with an incidence of 1 in 36,000 individuals. Tumors found in patients with VHL could be benign or malignant and depending on its type can include clear cell renal cell carcinoma (RCC), pheochromocytoma, hemangioblastoma of the cerebellum and spine, retinal angioma, endolymphatic sac tumors of the middle ear, pancreatic serous cystadenoma, and neuroendocrine tumors as well as papillary cystadenoma of the epididymis and broad ligament [45]. VHL type I confers a low risk of pheochromocytoma and VHL type II is associated with a high risk of pheochromocytoma, which is further subdivided into IIa, IIb, and IIc on the basis of the risk for developing renal cell cancer.

VHL disease is caused by mutations in the *VHL* gene, located on chromosome *3p25* which codes for pVHL, a tumor-suppressor protein. pVHL regulates levels of several proteins such as hypoxia-inducible factor-1 alpha (HIF-1α) and HIF-2α by targeting them for proteasomal degradation, and in VHL patients the levels of these proteins goes up, which further affects other factors such as

VEGF, platelet-derived growth factor (PDGF)-beta, and transforming growth factor (TGF)-alpha. These factors play a significant role in angiogenesis and tumorigenesis [45, 46].

Targeted therapy approaches for VHL with clinical activity are mostly agents with antiangiogenic properties such TKIs, mTOR inhibitors, and VEGF receptor blockers.

10.12.1 Tyrosine Kinase Inhibitors (TKIs)

Sunitinib is a multi-target tyrosine kinase inhibitor (PDGFR, VEGF, FMS-like tyrosine kinase-3, RET, CSF1R) which is FDA approved for the treatment of well-differentiated, progressive pancreatic neuroendocrine tumors (PNET) in patients with unresectable, locally advanced, or metastatic disease at a recommended dose is 37.5 mg daily and for advanced renal cell carcinoma at a dose of 50 mg once daily for 4 weeks of a 6-week treatment cycle. The clinical trial which led to approval of sunitinib for PNET only had two patients with VHL mutation. However, other smaller studies have shown clinical benefit of sunitinib in VHL patients [47]. In a pilot study which showed that sunitinib has activity against PNETs, no response was seen in hemangioblastomas associated with VHL [47]. No systemic targeted therapy has shown significant benefit in VHL-associated hemangioblastomas. The potential role of pazopanib was demonstrated in a case report [46]. Some other multikinase TKIs which have shown tolerability in phase 1 trials and are under phase 2 trials for VHL-associated hemangioblastomas are PTK787/ZK222584 (PTK/ZK, vatalanib) (NCI trial #NCT00052013) and TKI-538(NCI trial #NCT01266070).

Several TKIs have been widely studied as therapeutic agents for the treatment of VHL-associated RCC [47]. In a pilot study by Jonasch et al., 33% of RCCs showed a partial response, 67% were stable, and 10% progressed [47]. Multiple trials have been conducted to investigate the treatment of advanced clear cell renal cell carcinoma, most of which (80%) harbor mutations *VHL* gene. The TKIs studied in these trials include sorafenib, sunitinib, pazopanib, and axitinib, all of which have now been approved for the treatment of advanced clear cell RCC [48]. The role of these agents specifically in patients with VHL disease is less clear. It has been shown that patients with loss of functions (frameshift, nonsense, splice, and in-frame deletions/insertions) which are more characteristic of VHL disease have a better response rate to anti-VEGF therapy compared to wild-type VHL in patients with advanced RCC [49]. Sunitinib (Sutent) was being studied in a phase 2 clinical trial (NCI trial # NCT00330564) and preliminary results have showed partial response in 6 of 18 VHL renal lesions. The study was terminated due to slow accrual. Pazopanib is currently under clinical investigation in patients with VHL (NCI trial #NCT01436227 and NCT01436227).

10.12.2 mTOR Inhibitors

mTOR kinase regulates transcription and translation of HIF-1α (Fig. 10.1), which points towards a potential role of mTOR inhibitors in treating VHL patients [46]. Everolimus is approved for treatment of advanced, metastatic, or unresectable pancreatic neuroendocrine tumors (PNET) and advanced renal cell carcinoma, while temsirolimus is FDA approved for advanced renal cell carcinoma [48]. Although significant clinical activity was seen in these studies, the status of VHL mutation was not mentioned. However, most of the patients with clear cell-type RCC have VHL mutation, and there is a strong possibility, although not proven, that these agents are active in VHL syndrome.

10.12.3 Anti-VEGF Receptor Antibodies

Anti-VEGF receptor antibody bevacizumab and other similar agents have investigated tried for hemangioblastomas and RCC. Case reports of activity of systemic bevacizumab for hemangioblastomas (VHL status not known) and intravitreal bevacizumab for VHL-associated retinal hemangioblastomas can be found in literature. However, the benefit has not been proven in any larger study [46]. A clinical trial of bevacizumab was terminated early owing to low recruitment (NCI trial #NCT01015300). Bevacizumab is approved for treatment of advanced RCC, but as with other antiangiogenic agents mentioned earlier, the VHL mutation or disease status was not reported with the trial [50].

10.12.4 Other Agents Including Histone Deacetylases (HDAC) Inhibitor

Histone deacetylases (HDAC) inhibitor vorinostat is also in phase 1 clinical trial for nervous system hemangioblastomas in VHL patients with missense mutation only (NCI trial # NCT02108002).

10.13 Cowden Syndrome

Cowden syndrome or multiple hamartoma syndromes is inherited in an autosomal dominant fashion. It is a subtype of PTEN hamartoma tumor syndrome which manifests as dermatologic changes and multiple mucocutaneous or extracutaneous hamartomatous tumors involving multiple organs. Patients with germline *PTEN* mutations have an increased lifetime cumulative risk of developing breast cancer (77–85%), thyroid cancer (35–38%), endometrial cancer (28%), colorectal cancer (9%), renal cancer (34%), and melanoma (6%) [46, 51].

Patients with Cowden syndrome have germline mutations in the *PTEN* (phosphatase and tensin) homolog gene (25–35%). PTEN is a negative regulator of PI3K/AKT/mTOR signaling pathway and its loss seen in Cowden syndrome patients leads to activation of PI3K/AKT/mTOR signaling pathway (Fig. 10.1) [46]. Other mutations which have been reported in patients with Cowden syndrome include germline mutations in *PIK3CA* and *AKT1* [46, 52].

Agents targeting the PIK3CA/AKT/mTOR pathway may have a therapeutic role in these patients (Fig. 10.1) [46]. mTOR inhibitors have been tested in these patients and benefits have been reported in small group of patients with germline PTEN mutations [46, 53]. XL147, a pan PI3K inhibitor, showed more pronounced pathway modulation in hamartoma tumor tissue compared to the normal skin adjacent to tumor in a patient with Cowden syndrome, suggesting a possible role [54]. Clinical trials are evaluating the role of PIK3CA/mTOR inhibitors such as sirolimus (NCI trial #NCT00971789), BGT226 (NCI trial #NCT00600275), and BEZ235 (NCI trial #NCT00620594) for the treatment of malignancies in these patients. Selective PI3K beta inhibitors may have better activity in patients with Cowden syndrome, as it has been shown that inhibition of p110 beta is sufficient to inhibit tumor formation driven by PTEN loss [55]. Clinical trial (NCI trial # NCT01458067) is investigating the role of selective inhibitor of PI3Kβ in patients with PTEN deficient advanced tumors.

10.14 Proteus and Proteus-Like Syndrome

Proteus syndrome is not an inherited disorder like the other syndromes mentioned in this chapter; we have, however, included this topic due to similar concept and application of targeted therapy. It is a rare disorder comprising overgrowth and malformations of skin, connective tissue, fat, brain, and other tissues [46, 56]. Major clinical manifestations are irregular, progressive postnasal growth, cutaneous capillary, and venous malformations, monomorphic adenoma of the parotid glands, bilateral ovarian cystadenoma, unilateral cystadenoma, and meningioma [46, 56]. Clinic manifestations overlap with other overgrowth syndromes [56].

The AKT pathway is upregulated in Proteus syndrome due to underlying somatic activating mutations (c.49G→A, p.Glu17Lys) in the *AKT1* oncogene. AKT and mTOR inhibitors, which can impair the AKT pathway, may have a role in these patients [46, 53, 57, 58]. *PTEN* mutations have also been reported in patients with Proteus syndrome. Treatment with rapamycin did produce clinical benefit in a Proteus syndrome patient with a germline PTEN mutation [58].

10.15 Li–Fraumeni Syndrome

Li–Fraumeni syndrome is an autosomal dominant hereditary disorder with increased risk of early onset malignancies involving soft tissue, breast, lung, pancreas, skin, gastrointestinal tract, choroid plexus, and colon [59]. Patients are also at increased risk for lymphoma, melanoma, germ cell tumors, and Wilms tumor [17, 59].

The mutation in Li–Fraumeni involves the *TP53* gene, located on chromosome *17p13.1* [17, 59]. *TP53* is a tumor-suppressor gene that is involved in multiple cellular processes, including growth arrest, apoptosis, and DNA repair. These functions are impaired in Li–Fraumeni syndrome patients, leading to increase oncogeneity [17, 59, 60].

There is no specific agent with clinical activity in patients with Li-Fraumeni syndrome or patients with mutation in the *TP53* gene. A retrospective analysis did show a possible role of bevacizumab in patients with *TP53* mutation; however, there were limitations to that study and the results could be best considered hypothesis generating [61].

Other ways in which mutant *TP53* can be targeted have been suggested, such as increasing MDM2-mediated destruction of P53 by using histone deacetylase (HDAC) inhibitors [17, 60]. These strategies need to be explored clinically in patients with solid tumors who have Li–Fraumeni syndrome.

10.16 Conclusions

Understanding the genetics of hereditary cancer syndromes has also helped to better understand the genetic pathway in non-hereditary cancers. With the advancement of technologies such as next-gen sequencing, more number of germline mutations is being discovered. For instance, germline PTPRD mutations were recently reported in Ewing sarcoma, a malignancy that occurs most commonly in children, adolescents, and young adults [62, 63]. In addition, identification of germline mutations may be informative in orphan cancers [64]. So far, commercially available next generation sequencing panels have been available only for somatic aberrations, but more recently, germline aberration panels are also being made available for clinical use [9, 65–70]. This may open up several avenues for targeted therapy in these orphan cancers.

Development of targeted therapy also has faced several barriers. Rarity and poor participation in clinical trials is one of them. Development of cancer consortiums such as the neurofibroma clinical trials consortium can enhance patients' participation for such rare tumors [5]. Clinical trials should be designed to test for molecular aberrations, and correlations with response can help identify agents with potential benefits. Earlier molecular profiling in young patients with malignancies would be helpful in identifying possible underlying syndromes, allowing the effective utilization of targeted therapies in treatment as clinical trials progress.

Development of therapies has also been slow due because many of the genetic defects in hereditary syndrome involve tumor-suppressor genes (e.g., NF-1, NF-2, P53, TSC1/2) which are not targets, but the targets are downstream of the mutated protein, making it more complicated. Synthetic lethality as seen with PARP inhibitors and BRCA1/2 mutation has been a useful strategy. Combination therapy of targeted agents with another targeted agent and/or chemotherapy needs to be explored.

We have reviewed only a few of the cancer syndromes. With the commercial and clinical availability of germline panels, more patients may be diagnosed with aberrations predisposing them to cancer.

Disclosure

None of the authors have any relevant conflicts of interest to disclose.

Acknowledgments

VS thanks Shanon Wilkes osteosarcoma research foundation. The University of Texas MD Anderson Cancer Center is supported by National Institutes of Health Cancer Center Support Grant CA016672.

References

1. Agarwal, R., et al. Targeted therapy for hereditary cancer syndromes: neurofibromatosis type 1, neurofibromatosis type 2, and Gorlin syndrome. *Discov Med* **18**, 323–330 (2014).
2. Dilworth, J. T., et al. Molecular targets for emerging anti-tumor therapies for neurofibromatosis type 1. *Biochem Pharmacol* **72**, 1485–1492 (2006).

3. Widemann, B. C., et al. Phase 2 randomized, flexible crossover, double-blinded, placebo-controlled trial of the farnesyltransferase inhibitor tipifarnib in children and young adults with neurofibromatosis type 1 and progressive plexiform neurofibromas. *Neuro Oncol* **16**, 707–718 (2014).
4. Kim, A., et al. Phase I trial and pharmacokinetic study of sorafenib in children with neurofibromatosis type I and plexiform neurofibromas. *Pediatr Blood Cancer* **60**, 396–401 (2013).
5. Weiss, B., et al. Sirolimus for progressive neurofibromatosis type 1-associated plexiform neurofibromas: a Neurofibromatosis Clinical Trials Consortium phase II study. *Neuro Oncol* **17**, 596–603 (2015).
6. Robertson, K. A., et al. Imatinib mesylate for plexiform neurofibromas in patients with neurofibromatosis type 1: a phase 2 trial. *Lancet Oncol* **13**, 1218–1224 (2012).
7. Evans, D. G. Neurofibromatosis type 2 (NF2): a clinical and molecular review. *Orphanet J Rare Dis* **4**, 16 (2009).
8. Evans, D. G., et al. Consensus recommendations to accelerate clinical trials for neurofibromatosis type 2. *Clin Cancer Res* **15**, 5032–5039 (2009).
9. Subbiah, V., et al. Treatment of patients with advanced neurofibromatosis type 2 with novel molecularly targeted therapies: from bench to bedside. *J Clin Oncol* **30**, e64–e68 (2012).
10. Plotkin, S. R., et al. Hearing improvement after bevacizumab in patients with neurofibromatosis type 2. *N Engl J Med* **361**, 358–367 (2009).
11. Plotkin, S. R., et al. Erlotinib for progressive vestibular schwannoma in neurofibromatosis 2 patients. *Otol Neurotol* **31**, 1135–1143 (2010).
12. Karajannis, M. A., et al. Phase II trial of lapatinib in adult and pediatric patients with neurofibromatosis type 2 and progressive vestibular schwannomas. *Neuro Oncol* **14**, 1163–1170 (2012).
13. Lo Muzio, L. Nevoid basal cell carcinoma syndrome (Gorlin syndrome). *Orphanet J Rare Dis* **3**, 32 (2008).
14. Sekulic, A., et al. Efficacy and safety of vismodegib in advanced basal-cell carcinoma. *N Engl J Med* **366**, 2171–2179 (2012).
15. Tang, J. Y., and Marghoob, A. A. Emerging treatments and signaling pathway inhibitors. *Semin Cutan Med Surg* **30**, S14–S18 (2011).
16. Foulkes, W. D. Inherited susceptibility to common cancers. *N Engl J Med* **359**, 2143–2153 (2008).
17. Agarwal, R., et al. Targeted therapy for hereditary cancer syndromes: hereditary breast and ovarian cancer syndrome, Lynch syndrome,

familial adenomatous polyposis, and Li-Fraumeni syndrome. *Discov Med* **18**, 331–339 (2014).

18. Farmer, H., et al. Targeting the DNA repair defect in BRCA mutant cells as a therapeutic strategy. *Nature* **434**, 917–921 (2005).
19. Zhang, F., et al. PALB2 links BRCA1 and BRCA2 in the DNA-damage response. *Curr Biol* **19**, 524–529 (2009).
20. Kaufman, B., et al. Olaparib monotherapy in patients with advanced cancer and a germline BRCA1/2 mutation. *J Clin Oncol* **33**, 244–250 (2015).
21. Oza, A. M., et al. Olaparib combined with chemotherapy for recurrent platinum-sensitive ovarian cancer: a randomised phase 2 trial. *Lancet Oncol* **16**, 87–97 (2015).
22. Liu, J. F., et al. Combination cediranib and olaparib versus olaparib alone for women with recurrent platinum-sensitive ovarian cancer: a randomised phase 2 study. *Lancet Oncol* **15**, 1207–1214 (2014).
23. Isakoff, S., et al. A phase II trial expansion cohort of the PARP inhibitor veliparib (ABT888) and temozolomide in BRCA1/2 associated metastatic breast cancer. *Cancer Res* **71**, P3-16-05 (2012).
24. Gala, M., and Chung, D. C. Hereditary colon cancer syndromes. *Semin Oncol* **38**, 490–499 (2011).
25. Gupta, R. A., and Dubois, R. N. Colorectal cancer prevention and treatment by inhibition of cyclooxygenase-2. *Nat Rev Cancer* **1**, 11–21 (2001).
26. Burn, J., et al. Long-term effect of aspirin on cancer risk in carriers of hereditary colorectal cancer: an analysis from the CAPP2 randomised controlled trial. *Lancet* **378**, 2081–2087 (2011).
27. Narayan, S., and Roy, D. Role of APC and DNA mismatch repair genes in the development of colorectal cancers. *Mol Cancer* **2**, 41 (2003).
28. Giardiello, F. M., et al. Treatment of colonic and rectal adenomas with sulindac in familial adenomatous polyposis. *N Engl J Med* **328**, 1313–1316 (1993).
29. Steinbach, G., et al. The effect of celecoxib, a cyclooxygenase-2 inhibitor, in familial adenomatous polyposis. *N Engl J Med* **342**, 1946–1952 (2000).
30. Iwama, T., Akasu, T., Utsunomiya, J., and Muto, T. Does a selective cyclooxygenase-2 inhibitor (tiracoxib) induce clinically sufficient suppression of adenomas in patients with familial adenomatous polyposis? A randomized double-blind placebo-controlled clinical trial. *Int J Clin Oncol* **11**, 133–139 (2006).

31. van Es, J. H., et al. Notch/gamma-secretase inhibition turns proliferative cells in intestinal crypts and adenomas into goblet cells. *Nature* **435**, 959–963 (2005).
32. Mathew, C. G. Fanconi anaemia genes and susceptibility to cancer. *Oncogene* **25**, 5875–5884 (2006).
33. Agarwal, R., et al. Targeted therapy for genetic cancer syndromes: Fanconi anemia, medullary thyroid cancer, tuberous sclerosis, and RASopathies. *Discov Med* **19**, 101–108 (2015).
34. Villarroel, M. C., et al. Personalizing cancer treatment in the age of global genomic analyses: PALB2 gene mutations and the response to DNA damaging agents in pancreatic cancer. *Mol Cancer Ther* **10**, 3–8 (2011).
35. Frank-Raue, K., Rondot, S., and Raue, F. Molecular genetics and phenomics of RET mutations: impact on prognosis of MTC. *Mol Cell Endocrinol* **322**, 2–7 (2010).
36. Wells, S. A., Jr., et al. Vandetanib in patients with locally advanced or metastatic medullary thyroid cancer: a randomized, double-blind phase III trial. *J Clin Oncol* **30**, 134–141 (2012).
37. Elisei, R., et al. Cabozantinib in progressive medullary thyroid cancer. *J Clin Oncol* **31**, 3639–3646 (2013).
38. Lendvay, T. S., and Marshall, F. F. The tuberous sclerosis complex and its highly variable manifestations. *J Urol* **169**, 1635–1642 (2003).
39. Subbiah, V., Trent, J. C., and Kurzrock, R. Resistance to mammalian target of rapamycin inhibitor therapy in perivascular epithelioid cell tumors. *J Clin Oncol* **28**, e415 (2010).
40. Curatolo, P., Bombardieri, R., and Jozwiak, S. Tuberous sclerosis. *Lancet* **372**, 657–668 (2008).
41. Franz, D. N., et al. Efficacy and safety of everolimus for subependymal giant cell astrocytomas associated with tuberous sclerosis complex (EXIST-1): a multicentre, randomised, placebo-controlled phase 3 trial. *Lancet* **381**, 125–132 (2013).
42. Krueger, D. A., et al. Everolimus for subependymal giant-cell astrocytomas in tuberous sclerosis. *N Engl J Med* **363**, 1801–1811 (2010).
43. Tartaglia, M., Gelb, B. D., and Zenker, M. Noonan syndrome and clinically related disorders. *Best Pract Res Clin Endocrinol Metab* **25**, 161–179 (2011).
44. Tidyman, W. E., and Rauen, K. A. The RASopathies: developmental syndromes of Ras/MAPK pathway dysregulation. *Curr Opin Genet Dev* **19**, 230–236 (2009).

45. Maher, E. R., Neumann, H. P., and Richard, S. von Hippel-Lindau disease: a clinical and scientific review. *Eur J Hum Genet* **19**, 617–623 (2011).
46. Agarwal, R., et al. Targeted therapy for genetic cancer syndromes: von Hippel-Lindau disease, Cowden syndrome, and Proteus syndrome. *Discov Med* **19**, 109–116 (2015).
47. Jonasch, E., et al. Pilot trial of sunitinib therapy in patients with von Hippel-Lindau disease. *Ann Oncol* **22**, 2661–2666 (2011).
48. Randall, J. M., Millard, F., and Kurzrock, R. Molecular aberrations, targeted therapy, and renal cell carcinoma: current state-of-the-art. *Cancer Metastasis Rev* **33**, 1109–1124 (2014).
49. Choueiri, T. K., et al. von Hippel-Lindau gene status and response to vascular endothelial growth factor targeted therapy for metastatic clear cell renal cell carcinoma. *J Urol* **180**, 860–865; discussion 865–866 (2008).
50. Escudier, B., et al. Bevacizumab plus interferon alfa-2a for treatment of metastatic renal cell carcinoma: a randomised, double-blind phase III trial. *Lancet* **370**, 2103–2111 (2007).
51. Tan, M. H., et al. Lifetime cancer risks in individuals with germline PTEN mutations. *Clin Cancer Res* **18**, 400–407 (2012).
52. Orloff, M. S., et al. Germline PIK3CA and AKT1 mutations in Cowden and Cowden-like syndromes. *Am J Hum Genet* **92**, 76–80 (2013).
53. Munoz, J., and Kurzrock, R. Targeted therapy in rare cancers–adopting the orphans. *Nat Rev Clin Oncol* **9**, 631–642 (2012).
54. Shapiro, G. I., et al. Phase I safety, pharmacokinetic, and pharmacodynamic study of SAR245408 (XL147), an oral pan-class I PI3K inhibitor, in patients with advanced solid tumors. *Clin Cancer Res* **20**, 233–245 (2014).
55. Ni, J., et al. Functional characterization of an isoform-selective inhibitor of PI3K-p110beta as a potential anticancer agent. *Cancer Discov* **2**, 425–433 (2012).
56. Cohen, M. M., Jr. Proteus syndrome: an update. *Am J Med Genet C Semin Med Genet* **137C**, 38–52 (2005).
57. Lindhurst, M. J., et al. A mosaic activating mutation in AKT1 associated with the Proteus syndrome. *N Engl J Med* **365**, 611–619 (2011).
58. Marsh, D. J., et al. Rapamycin treatment for a child with germline PTEN mutation. *Nat Clin Pract Oncol* **5**, 357–361 (2008).
59. Malkin, D. Li-fraumeni syndrome. *Genes Cancer* **2**, 475–484 (2011).
60. Muller, P. A., and Vousden, K. H. p53 mutations in cancer. *Nat Cell Biol* **15**, 2–8 (2013).

61. Said, R., et al. P53 mutations in advanced cancers: clinical characteristics, outcomes, and correlation between progression-free survival and bevacizumab-containing therapy. *Oncotarget* **4**, 705–714 (2013).
62. Jiang, Y., et al. Germline PTPRD mutations in Ewing sarcoma: biologic and clinical implications. *Oncotarget* **4**, 884–889 (2013).
63. Subbiah, V., et al. Morphoproteomic profiling of the mammalian target of rapamycin (mTOR) signaling pathway in desmoplastic small round cell tumor (EWS/WT1), Ewing's sarcoma (EWS/FLI1) and Wilms' tumor(WT1). *PLoS One* **8**, e68985 (2013).
64. Subbiah, V., et al. Bilateral gonadoblastoma with dysgerminoma and pilocytic astrocytoma with WT1 GT-IVS9 mutation: a 46 XY phenotypic female with Frasier syndrome. *Pediatr Blood Cancer* **53**, 1349–1351 (2009).
65. Jardim, D. L., Conley, A., and Subbiah, V. Comprehensive characterization of malignant phyllodes tumor by whole genomic and proteomic analysis: biological implications for targeted therapy opportunities. *Orphanet J Rare Dis* **8**, 112 (2013).
66. Subbiah, V., et al. Targeted therapy by combined inhibition of the RAF and mTOR kinases in malignant spindle cell neoplasm harboring the KIAA1549-BRAF fusion protein. *J Hematol Oncol* **7**, 8 (2014).
67. Subbiah, V. Prospects and pitfalls of personalizing therapies for sarcomas: from children, adolescents, and young adults to the elderly. *Curr Oncol Rep* **16**, 401 (2014).
68. Subbiah, V., et al. Next generation sequencing analysis of platinum refractory advanced germ cell tumor sensitive to Sunitinib (Sutent(R)) a VEGFR2/PDGFRbeta/c-kit/ FLT3/RET/CSF1R inhibitor in a phase II trial. *J Hematol Oncol* **7**, 52 (2014).
69. Egas-Bejar, D., et al. Theranostic profiling for actionable aberrations in advanced high risk osteosarcoma with aggressive biology reveals high molecular diversity: the human fingerprint hypothesis. *Oncoscience* **1**, 167–179 (2014).
70. Jiang, Y., et al. Novel secondary somatic mutations in Ewing's sarcoma and desmoplastic small round cell tumors. *PLoS One* **9**, e93676 (2014).

Chapter 11

Pathology in the Era of Personalized Medicine

Hye Seung Lee
Department of Pathology, Seoul National University Bundang Hospital, 173-82 Gumi-ro, Bundang-gu, Seongnam-si, Gyeonggi-do, 13620, Republic of Korea
hye2@snu.ac.kr

To manage cancer patients in daily practice, all that had been required of pathologists was an accurate and timely pathologic diagnosis. However, since our understanding of the genetic changes that lead to carcinogenesis, cancer progression, and metastasis has improved [1] and targeted therapies based on these genetic events have been developed for various cancers [2], the focus has shifted towards the use of molecular analyses to identify patients that are most likely to respond to these therapies. Therefore, in addition to accurate and timely diagnoses, complete reporting that can guide patient management is needed from pathologists, and as such, they will play an integral role in personalized medicine. In this chapter, the role and clinical significance of molecular pathology in personalized medicine are covered.

Cancer Genetics and Genomics for Personalized Medicine
Edited by Il-Jin Kim

ISBN 978-981-4669-87-0 (Hardcover), 978-1-315-36476-6 (eBook)
www.panstanford.com

11.1 Why Is the Role of Pathologists in Personalized Medicine Important?

With the recent developments in cancer pathology, a pathologic diagnosis needs to include molecular and genomic information in addition to cancer grading, staging, and histologic findings with prognostic significance. Recently applied targeted cancer therapies demand that pathologists develop and use suitable detection techniques, variously termed "predictive markers," "companion diagnostics," or "advanced personalized diagnostics" [3]. An error in cancer pathology may be a miss or misinterpretation of an important predictive or prognostic genetic finding.

Today, diagnosis of metastatic colorectal carcinoma is a full-page report containing detailed information on grading, staging, and predictive and prognostic markers, such as microsatellite instability and *RAS* mutation status [4, 5]. The molecular pathologic diagnosis should be reliable and reproducible, and the pathologists will have primary responsibility for the molecular diagnosis in personalized medicine.

Personalized medicine requires multidisciplinary coordination of the surgical, oncology, research, and pathology teams to ensure an accurate and timely diagnosis. Personalized medicine demands personalized pathology with the use of molecular and genomic techniques. Pathologists can play a central role in this process by selecting and preparing tissue samples, selecting genetic assays, and by standardizing the reporting of results. Molecular diagnoses are now accepted in various cancers, such as HER2 status in breast and gastric carcinoma, *EGFR* mutations and *ALK* rearrangements in lung cancer, *BRAF* mutations in malignant melanoma, and *RAS* mutations in colorectal cancer [6]. These molecular techniques offer some advantages; however, they are still evolving. Therefore, preanalytic, analytic, and postanalytic considerations are required for personalized pathology (Fig. 11.1).

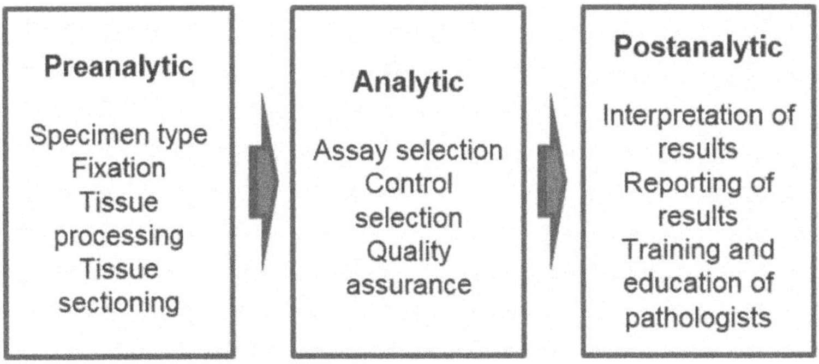

Figure 11.1 Standardization of molecular pathology.

11.2 Practical Guidance for Molecular Pathology

11.2.1 Preanalytic

Sampling is an early determinant of the accuracy of molecular testing. Biopsy specimens are more commonly used than surgical resection specimens because patients with advanced and inoperable cancer usually become candidates for targeted therapy. However, heterogeneity in genetic status has been reported for various cancers. For example, heterogeneous staining of HER2 is common in gastric carcinoma and is estimated to be present in 10–40% of IHC 3+ cases and 60–95% in IHC 2+ cases [7–9]. Kosmidou et al. reported that differences in *KRAS* mutations between the tumor center and periphery occur in 50.7% of colorectal cancer specimens [10]. Therefore, biopsy specimens should be taken carefully to acquire sufficient viable tumor samples.

The accuracy and reproducibility of molecular diagnosis depend on the quantity and quality of the cancer tissue specimens. Although snap freezing of tumor tissues in liquid nitrogen is the optimal method for preserving nucleic acids, in many pathologic laboratories, tumor tissues usually are formalin-fixed and paraffin-embedded (FFPE) to preserve histological features. The nucleic acids and proteins in tissue specimens are affected by the type of fixative used, the duration of fixation, decalcification, and the storage conditions (i.e., time, temperature, and humidity) [11]. The recommended method is to quickly transfer the specimen into 10% neutral buffered formalin, usually within 1 hour, and to fix the specimen for 8–72 hours [6, 12]. Both prolonged preanalytic cold ischemia time and overfixation or underfixation can cause false-negative or invalid results. In addition, the storage conditions and duration of storage for the FFPE blocks can affect test results. Even if the tumor tissue is frozen, the nucleic acids degenerate and become fragmented with after several years of storage [11].

Pathologists verify the presence of tumor cells in the tissue section or block to be used for genetic testing and assess the quantity and quality of tumor material [13]. Freezing tumor samples are usually taken by the gross findings. The center of gastric and colorectal cancer is usually inflamed owing to the presence of ulcers, and the periphery contains a mix of tumor and normal tissue; therefore, it is not easy to collect an area containing tumor only. In addition, in breast and prostate cancer, it is difficult to discriminate the tumor area by gross examination. Therefore, pathologists determine whether the percentage of the tumor area in the tissue sample is above the lower limit of quantitation for genetic testing by microscopic examination using frozen tissue or FFPE tissues. In the tumor area, there is a variable amount of non-neoplastic cells and materials, including inflammatory cells, fibroblasts, endothelial cells, and mucin (Fig. 11.2). Therefore, microdissection is sometimes required. In addition, pathologists have to keep in mind that adenoma and noninvasive carcinomas should not be used for molecular testing [14].

11.2.2 Analytic

Regarding the selection of a molecular test for targeted therapies, standardized and commercially available assays are preferred over laboratory-developed assays, especially for immunohistochemistry (IHC) and in situ hybridization (ISH). A decade ago, herceptin was administered to patients with human epidermal growth factor receptor-2 (HER2)-positive cancer, including breast and gastric carcinomas [15, 16]. Now, it is more important for pathologists to inform oncologists of HER2 status than the histologic subtype. Validated, commercially available methodologies are recommended

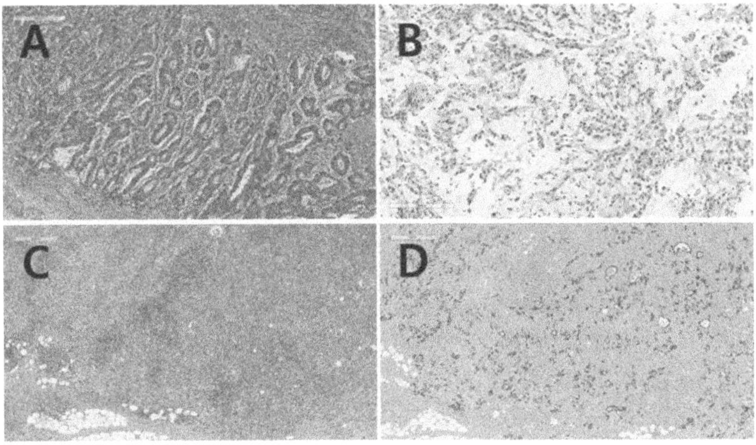

Figure 11.2 Histologic features of tumor area in various subtype of gastric carcinoma. Quantity of tumor cells in grossly tumor area is various according to subtype. (A) Carcinoma cells consist of most tumor area in intestinal type. (B) Extracellular mucin pool is observed in about a half of tumor area. (C) Rich infiltration of lymphoid cells is found in gastric carcinoma with lymphoid stroma. (D) Epstein–Barr virus (EBV) in situ hybridization of C. EBV-positive carcinoma cells are observed in less than a third of the tumor area.

for accurate and reliable HER2 results. Ideally, assays should be validated and approved by regional regulatory agencies [12, 14].

For mutation analyses, many different methodologies can be used, including Sanger sequencing, commercial testing kits, pyrosequencing, next-generation sequencing (NGS), and other laboratory-developed tests [17]. Mutations in the *RAS* gene predict a lack of response to panitumumab and cetuximab in patients with metastatic colorectal carcinoma [5]. Each method for detecting *KRAS* mutations has its advantages and disadvantages. Although Sanger sequencing is considered the gold standard method, its sensitivity is low and it is a labor-intensive and time-consuming method. In contrast, commercial kits are rapid, simple, and very sensitive. However, they are expensive and cannot detect all mutations [17]. Therefore, each method requires validation before implementation in the laboratory to ensure the accuracy of the reported results.

11.2.3 Quality Assurance

When establishing diagnostic testing for targeted therapy in any center, initial test validation is required. For HER2 testing in gastric carcinoma, the first 25–50 cases should be analyzed in parallel using both IHC and ISH. The concordance rate should be >90% if the equivocal IHC 2+ group is disregarded [12]. It should be done by personnel specifically trained in HER2 testing methods and interpretation for gastric carcinoma, irrespective of their previous experience in breast carcinoma. For mutation analysis, the initial validation test should be done using both samples with known mutational status and tissue specimens with unknown status processed in each institute [12, 14].

All laboratories providing molecular pathology services should actively participate in proficiency testing schemes through external quality assurance programs, such as ISO 15189 or other national equivalent [6]. In addition, laboratories should have their own internal quality assurance program. The relationship between the results and clinicopathologic features, concordance between methods, and positivity rates needs to be monitored regularly. Furthermore, the use of control materials is essential for all molecular tests. Both negative and positive controls should be used in each IHC, ISH, and mutation assay, and the results of internal control testing should be monitored.

In order to achieve qualified and consistent testing results in light of the lack of new equipment and properly trained pathologists, oncologists have attempted to request molecular testing at central laboratories. Central laboratories have several advantages, including sufficient experience, governance by a validated procedure, and continuous performance of adequate numbers of molecular tests to maintain quality standards. However, testing by central laboratories also has several disadvantages; for example, the optimal conditions for a molecular test differ among hospitals because of differences in sample quality and the procedures used for tissue processing, but a central laboratory cannot perform the tests using the optimal conditions for each hospital. In addition, it is more difficult for clinicians to communicate with pathologists located at central laboratories.

11.2.4 Postanalytic

A pathology report should accurately convey all the information a clinician needs to treat the patient, in sufficient detail to allow correct interpretation of the results. Therefore, a formatted diagnosis is required that contains the patient identification, results, test performed, any limitations of the tests, information about fixation, the individual taking responsibility, and the interpretation. In addition, integrated reporting of results from histopathology, IHC, ISH, and multiple genetic tests is also required. However, this is difficult to achieve in daily practice due to differences in reporting times and systems. Communication between pathologists and clinicians is very important for integration of this information and optimal personalized therapy.

The diagnostic criteria and interpretation are not the same for each tumor. For example, the interpretation and scoring guidelines for HER2 status in breast and gastric carcinoma differ. Unlike breast carcinoma, gastric carcinoma is a gland-forming, mucin-producing carcinoma that shows incomplete basolateral/lateral staining patterns in HER2 IHC. HER2 amplification is defined as a *HER2/CEP17* ratio $\geq$2.0 in gastric carcinoma, whereas is it defined as a *HER2/CEP17* ratio $\geq$2.0 or an average *HER2* copy number $\geq$6.0 signals/cell in breast carcinoma [14, 18]. For this reason, pathologists with subspecialty expertise are necessary, and they can communicate with clinical subspecialty groups to provide personalized care to cancer patients [19]. As a general surgical pathologist cannot be an expert on all organ systems.

Pathologist training should meet the accreditation requirements in each country. The staff responsible for the molecular pathology must have sufficient knowledge and experience in the preanalytic, analytic, and postanalytic pathways, and the molecular pathology reports need to be authorized by the staff that are responsible for diagnoses based on tissue and cytological material. In addition, to ensure safe and effective reporting of test results, it is essential that pathologists receive regular opportunities for continuous professional development. All groups involved, including scientists, oncologists, and surgeons must be taught how to communicate the molecular information and select the appropriate treatment [20].

11.3 Next-Generation Sequencing and the Pathologist

Among the recently emerged techniques, NGS has had perhaps the greatest impact [21]. The increased speed, efficiency, and resolution of NGS have greatly facilitated the detection of genetic, genomic, and epigenomic alterations. Whole genome sequencing of lung adenocarcinoma tissue from a young male non-smoker identified a novel rearrangement involving kinesin family member 5B and RET proto-oncogene (*KIF5B-RET*) that results in overexpression of the chimeric RET receptor, suggesting a promising target for personalized therapy [22]. In addition to identifying rare target genes, NGS can be used for targeted sequence enrichment to selectively capture genomic regions of interest, as an approach that could overcome therapeutic resistance, and a very high-resolution diagnostic tool to overcome the noise of tumor heterogeneity [23]. NGS technologies have also contributed to our understanding of the pathology of cancer through a comprehensive analysis of cancer genomes, including whole genomes, exomes, and transcriptomes.

These comprehensive analyses are possible using fresh tissue, body fluids, blood samples, or FFPE samples. To effectively use NGS technology in routine pathologic diagnosis, each step needs to be standardized to generate reproducible results. Similar to other molecular diagnostic tests, the preanalytic, analytic, and postanalytic workflow should be standardized and qualified. In particular, the pathologists are responsible for specimen handling and determining whether the specimen quality and percentage of cancer cells are acceptable for further NGS analysis. Recently, high correlation of DNA-exome and RNA sequence data between frozen samples and paired FFPE samples has been reported [24]. However, when using partially degraded DNA from FFPE tissues, additional technical issues are raised because of variations in tissue processing and storage. Therefore, an effective standard workflow for qualification and quantification of DNA preparation is needed when using FFPE tissues, considering the expensive and labor-intensive nature of NGS analysis [25]. Pathologists should understand the complexity of the testing procedures as well as the interpretation of NGS test results and actively participate in the NGS test team.

11.4 Conclusions

Recent advances in molecular and genomic techniques have enabled us to treat individual cancer patients with targeted and personalized therapy. Pathologists need to reliably provide molecular and genomic information to the clinician, and the pathology laboratory should have their own quality assurance program and be authorized by a regional or international agency in terms of the preanalytic, analytic, and postanalytic processes. In particular, pathologists must determine the specimen quality and authorize the test results. Active adaptation and innovation is required of pathologists in the personalized medicine era. Additionally, the clinician needs to communicate with the pathologists when conducting targeted therapy and clinical trials.

References

1. Hanahan D, Weinberg RA. Hallmarks of cancer: the next generation. *Cell.* 2011;144(5):646–674.
2. Dancey JE, Bedard PL, Onetto N, Hudson TJ. The genetic basis for cancer treatment decisions. *Cell.* 2012;148(3):409–420.
3. US Foodand Drug Administration. *Draft Guidance for Industry and Food and Drug Administration Staff-Invitro Companion Diagnostic Devices* (2011). Available from: http://www.fda.gov/MedicalDevices/deviceregulationandguidance/guidancedocuments/default.htm
4. Popat S, Hubner R, Houlston RS. Systematic review of microsatellite instability and colorectal cancerprognosis. *J Clin Oncol.* 2005;23(3):609–618.
5. Misale S, Di Nicolantonio F, Sartore-Bianchi A, Siena S, Bardelli A. Resistance to anti-EGFR therapy in colorectal cancer: from heterogeneity to convergent evolution. *Cancer Discov.* 2014;4(11):1269–1280.
6. Cree IA, Deans Z, Ligtenberg MJ, Normanno N, Edsjö A, Rouleau E, Solé F, Thunnissen E, Timens W, Schuuring E, Dequeker E, Murray S, Dietel M,

Groenen P, Van Krieken JH; European Society of Pathology Task Force on Quality Assurance in Molecular Pathology; Royal College of Pathologists. Guidance for laboratories performing molecular pathology for cancer patients. *J Clin Pathol.* 2014;67(11):923–931.

7. Kim KC, Koh YW, Chang HM, Kim TH, Yook JH, Kim BS, Jang SJ, Park YS. Evaluation of HER2 protein expression in gastric carcinomas: comparative analysis of 1,414 cases of whole-tissue sections and 595 cases of tissue microarrays. *Ann Surg Oncol.* 2011;18(10):2833–2840.
8. Lee HE, Park KU, Yoo SB, Nam SK, Park do J, Kim HH, Lee HS. Clinical significance of intratumoral HER2 heterogeneity in gastric cancer. *Eur J Cancer.* 2013;49(6):1448–1457.
9. Cho J, Jeong J, Sung J, Sung CO, Kim KM, Park CK, Choi MG, Sohn TS, Bae JM, Kim S. A large cohort of consecutive patients confirmed frequent HER2 positivity in gastric carcinomas with advanced stages. *Ann Surg Oncol.* 2013;20(Suppl 3):S477–S484.
10. Kosmidou V, Oikonomou E, Vlassi M, Avlonitis S, Katseli A, Tsipras I, Mourtzoukou D, Kontogeorgos G, Zografos G, Pintzas A. Tumor heterogeneity revealed by KRAS, BRAF, and PIK3CA pyrosequencing: KRAS and PIK3CA intratumor mutation profile differences and their therapeutic implications. *Hum Mutat.* 2014;35(3):329–340.
11. Nam SK, Im J, Kwak Y, Han N, Nam KH, Seo AN, Lee HS. Effects of fixation and storage of human tissue samples on nucleic acid preservation. *Korean J Pathol.* 2014;48(1):36–42.
12. Rüschoff J, Hanna W, Bilous M, Hofmann M, Osamura RY, Penault-Llorca F, van de Vijver M, Viale G. HER2 testing in gastric cancer: a practical approach. *Mod Pathol.* 2012;25(5):637–650.
13. Dijkstra JR, Heideman DA, Meijer GA, Boers JE, 't Hart NA, Diebold J, Hirschmann A, Hoefler G, Winter G, Miltenberger-Miltenyi G, Pereira SV, Richman SD, Quirke P, Rouleau EL, Guinebretiere JM, Tejpar S, Biesmans B, van Krieken JH. KRAS mutation analysis on low percentage of colon cancer cells: the importance of quality assurance. *Virchows Arch.* 2013;462(1):39–46.
14. Kim KM, Bilous M, Chu KM, Kim BS, Kim WH, Park YS, Ryu MH, Sheng W, Wang J, Chao Y, Ying J, Zhang S. Human epidermal growth factor receptor 2 testing in gastric cancer: recommendations of an Asia-Pacific task force. *Asia Pac J Clin Oncol.* 2014;10(4):297–307.
15. Slamon DJ, Leyland-Jones B, Shak S, Fuchs H, Paton V, Bajamonde A, Fleming T, Eiermann W, Wolter J, Pegram M, Baselga J, Norton L. Use of chemotherapy plus a monoclonal antibody against HER2

for metastatic breast cancer that overexpresses HER2. *N Engl J Med.* 2001;344(11):783–792.

16. Bang YJ, Van Cutsem E, Feyereislova A, Chung HC, Shen L, Sawaki A, Lordick F, Ohtsu A, Omuro Y, Satoh T, Aprile G, Kulikov E, Hill J, Lehle M, Rüschoff J, Kang YK; ToGA Trial Investigators. Trastuzumab in combination with chemotherapy versus chemotherapy alone for treatment of HER2-positive advanced gastric or gastro-oesophageal junction cancer (ToGA): a phase 3, open-label, randomised controlled trial. *Lancet.* 2010;376(9742):687–697.
17. Herreros-Villanueva M, Chen CC, Yuan SS, Liu TC, Er TK. KRAS mutations: analytical considerations. *Clin Chim Acta.* 2014;431:211–220.
18. Wolff AC, Hammond ME, Hicks DG, Dowsett M, McShane LM, Allison KH, Allred DC, Bartlett JM, Bilous M, Fitzgibbons P, Hanna W, Jenkins RB, Mangu PB, Paik S, Perez EA, Press MF, Spears PA, Vance GH, Viale G, Hayes DF; American Society of Clinical Oncology; College of American Pathologists. Recommendations for human epidermal growth factor receptor 2 testing inbreast cancer: American Society of Clinical Oncology/College of American Pathologists clinical practice guideline update. *J Clin Oncol.* 2013;31(31):3997–4013.
19. Black-Schaffer WS, Young RH, Harris NL. Subspecialization of surgical pathology at the Massachusetts General Hospital. *Am J Clin Pathol.* 1996;106(4 Suppl 1):S33–S42.
20. Tembuyser L, Ligtenberg MJ, Normanno N, Delen S, van Krieken JH, Dequeker EM. Higher quality of molecular testing, an unfulfilled priority: Results from external quality assessment for KRAS mutation testing in colorectal cancer. *J Mol Diagn.* 2014;16(3):371–377.
21. Meyerson M, Gabriel S, Getz G. Advances in understanding cancer genomes through second-generation sequencing. *Nat Rev Genet.* 2010;11(10):685–696.
22. Ju YS, Lee WC, Shin JY, Lee S, Bleazard T, Won JK, Kim YT, Kim JI, Kang JH, Seo JS. A transforming KIF5B and RET gene fusion in lung adenocarcinoma revealed from whole-genome and transcriptome sequencing. *Genome Res.* 2012;22(3):436-445.
23. Coco S, Truini A, Vanni I, Dal Bello MG, Alama A, Rijavec E, Genova C, Barletta G, Sini C, Burrafato G, Biello F, Boccardo F, Grossi F. Next generation sequencing in non-small cell lung cancer: new avenues toward the personalized medicine. *Curr Drug Targets.* 2015;16(1):47–59.

24. Hedegaard J, Thorsen K, Lund MK, Hein AM, Hamilton-Dutoit SJ, Vang S, Nordentoft I, Birkenkamp-Demtröder K, Kruhøffer M, Hager H, Knudsen B, Andersen CL, Sørensen KD, Pedersen JS, Ørntoft TF, Dyrskjøt L. Next-generation sequencing of RNA and DNA isolated from paired fresh-frozen and formalin-fixed paraffin-embedded samples of human cancer and normal tissue. *PLoS One*. 2014 May 30;9(5):e98187.

25. Simbolo M, Gottardi M, Corbo V, Fassan M, Mafficini A, Malpeli G, Lawlor RT, Scarpa A. DNA qualification workflow for next generation sequencing of histopathological samples. *PLoS One*. 2013;8(6):e62692.

Chapter 12

MicroRNAs in Human Cancers

Tae Jin Lee and Carlo M. Croce

Department of Cancer Biology and Genetics, Comprehensive Cancer Center, The Ohio State University, 1060 Biomedical Research Tower, 460 W. 12th Ave., Columbus, OH 43210, USA

taejin.lee@osumc.edu, carlo.croce@osumc.edu

12.1 Introduction

Human cancer is the result of genetic modifications and epigenetic alterations. During the last two decades, numerous findings that have been accumulated show that those genetic and epigenetic alterations are not limited to protein coding genes. Rather, other species of genetic information, called non-coding RNAs (ncRNA), arose as other major players that can deliver genetic information into biological phenotypes together with proteins. NcRNAs are defined as transcribed RNA species that are not further translated into proteins since they lack open reading frames (ORFs), which consist of start (AUG) and stop (UAA, UGA, or UAG) codons. Even though they are not used as protein-coding templates, ncRNAs are found to be biologically functional and involving a number of biological processes [1]. As the list of ncRNAs grows year by

Cancer Genetics and Genomics for Personalized Medicine
Edited by Il-Jin Kim

ISBN 978-981-4669-87-0 (Hardcover), 978-1-315-36476-6 (eBook)
www.panstanford.com

year, ncRNAs are now divided into two major categories simply on the basis of their transcript sizes: (i) RNAs longer than 200 nucleotides are long non-coding RNAs (lncRNAs), and (ii) RNAs shorter than 200 nucleotides are short non-coding RNAs (sncRNAs), which include microRNAs (miRNAs) and which this chapter focuses on. Such classification has been well reviewed by Gibb et al. [2]. First lncRNA, mouse H19, was discovered in 1990 [3], earlier than first miRNA, Lin-4 from *C. elegans* in 1993 [4]. However, lncRNA research has not been extensively explored until the last decade. Since the first discovery of miRNA involvement in human cancer in 2002 [5] that has exploded miRNA research linking to human diseases, several lncRNAs have been also implicated in human cancers. For example, H19 has been proposed as an oncogenic lncRNA since its expression was highly upregulated by c-Myc in multiple human cancers including liver, colon and breast [6] but downregulated by tumor suppressor p53 [7]. HOX antisense intergenic RNA (HOTAIR) was the first lncRNA that provided a direct link between lncRNA and metastatic breast cancer as upregulated HOTAIRs associate with polycomb repressive complex 2 (PRC2) proteins to downregulate a tumor suppressor HOXD, resulting increased cancer invasiveness and metastasis [8]. X-inactive specific transcript (XIST), found to be critical for X-chromosome inactivation in 1991 [9], expression correlates with tumor developments although its critical role in tumorigenesis needs to be elucidated further [10]. Metastasis-associated lung adenocarcinoma transcript 1 (MALAT1) was first discovered to be highly associated with metastatic potential and poor prognosis in non-small cell lung cancers [11] and later found to be highly upregulated in many cancers, including pancreatic [12], esophageal [13], renal [14], and colorectal [15] cancers. In contrast to these oncogenic lncRNAs, maternally expressed gene 3 (MEG3) was the first lncRNA proposed as a tumor suppressor in pituitary adenoma [16]. Its decreased expression and anti-tumor function were confirmed in many cancers including brain, gastric, liver and bladder cancers [17–20]. As of 2015, global transcriptome sequencing and mapping projects, such as FANTOM (Functional Annotation of Mammalian cDNA), predicted the existence of at least 35,000 ncRNAs in human, indicating there may be more lncRNAs than protein coding genes. However, the whole transcriptome analysis by cDNA sequencing and sequence annotation has a limitation to distinguish lncRNAs from mRNAs. Therefore, it is highly likely that the overall number of functional lncRNAs may increase as profiling technique advances in the near future.

MicroRNAs (miRNAs) are smallest member among all known ncRNAs as they are about 22 nucleotides in length. The existence of miRNAs was first uncovered in *C. elegance* in 1993 by two independent laboratories [4, 21]. Both studies identified a small piece of RNA transcribed from gene Lin4 regulating the expression of gene Lin14 through sequence complementarity. These remarkable observations, however, had not been largely recognized in medical biology field for several years until the discovery of short interfering RNA (siRNA) in 1999. Posttranscriptional gene silencing (PTGS) activity by small antisense RNA was first described in plants [22] and such RNA interfering (RNAi) was successfully demonstrated in mammalian cells with synthetic artificial siRNAs in 2001 [23]. The serious discoveries, Nobel Prize awarded in 2006, brought a great attention to the researchers who work in biomedical engineering and drug development for human gene therapy. It also shed a light on miRNA once more, since both seemed to resemble to each other from their small sizes to working mechanisms by RNA complementarity. Re-started in 2001, a number of small non-coding RNAs with gene-regulating roles were found in *C. elegance* and finally named as miRNAs [24–26]. In the following years, a vast number of miRNAs were identified in almost every eukaryotes, including human. Most recent version of miRBase, a handy miRNA database established by University of Manchester, UK, released in June 2014, annotated 28645 entries representing hairpin precursor miRNAs, expressing 35828 mature miRNA products, in 223 species. Up to date, there are 2,588 miRNAs in human and the number is growing. David Bartel laboratory in Massachusetts Institute of Technology (MIT) has estimated that miRNAs regulate more than 60% of human genes involved in biological processes [27], including cell cycle [28], differentiation and development [29, 30], metabolism [31], and human diseases, such as diabetes [32], cardiovascular diseases [33], neurodegenerative disorders [34, 35], and even cancers [5, 36]. In 2002, Carlo Croce laboratory found that miR-15/16 cluster is frequently deleted in chronic lymphocytic leukemia (CLL) patients, and those miRNAs directly regulate oncogene BCL2 to induce apoptosis [5]. This finding was the first pathological evidence that directly links miRNAs to human cancer development. The last decade was an era of miRNAs in human cancers, since numerous findings showing specific targeting of coding genes by specific miRNAs have been reported in various human cancers.

12.2 Biogenesis and Working Mechanism of MicroRNA

MiRNA coding genes are located either within the introns or exons of protein-coding genes (70%) or in intergenic areas (30%), and mostly found to be evolutionarily conserved [27]. Intragenic miRNAs are often parallel to their host gene, suggesting those miRNAs may be transcribed together with their host gene, while intergenic miRNA genes possess their own independent transcription units. Most miRNAs are initially transcribed into a long primary transcript, often up to several kilobases, by polymerase II [37] (Fig. 12.1). The long primary transcript, called pri-miRNA, is immediately processed into a shorter hairpin-like pre-miRNA in the nucleus by nuclear RNase III, called Drosha [38] Pre-miRNAs, 70 nucleotides long, are exported to cytoplasm by Exportin-5 [39], where it is further cleaved by a cytoplasmic RNase III, called Dicer, to generate 22 nucleotides of mature miRNA by removing its terminal loop [40, 41]. Then, the mature miRNA binds to Argonaute 2 (Ago2) and transactivation-responsive RNA-binding protein (TRBP) to form the RNA-induced silencing complex (RISC). RISC-associated miRNA is now active to function on its target mRNA. Of the two strands in mature miRNAs, active strand is retained in RISC while the other strand is degraded. The mature strand in RISC is guided to the 3′ untranslated region (3′ UTR) of target mRNAs by recognizing complementary sequences. The sequences of mature strand between the second and the eighth nucleotide, called "seed sequences," are known to be responsible for the complementarity, but partially. The partial base pairing between the seed sequences of miRNA and the 3′ UTR of target mRNA abolishes mechanistic action of translational

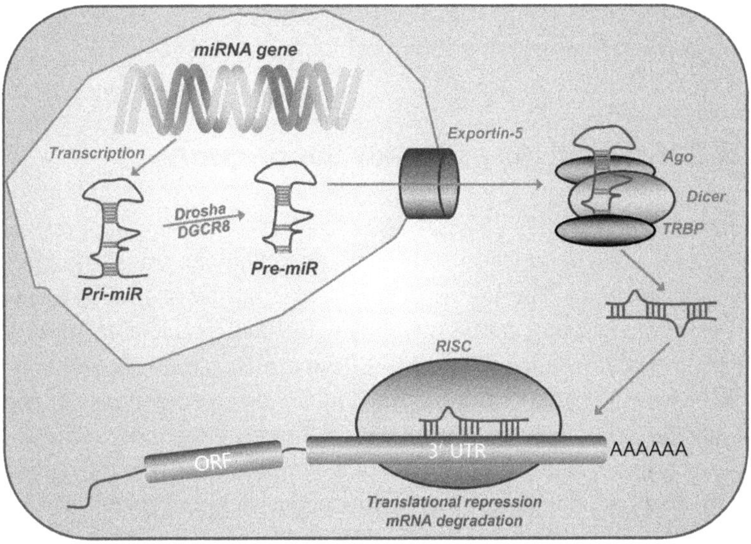

Figure 12.1 Biogenesis of microRNAs (miRNAs) involves multistep processes that can be divided into transcription, nuclear cropping, export to the cytoplasm, and cytoplasmic dicing. MicroRNA genes are transcribed into long, capped, and polyadenylated primary miRNAs (pri-miRNAs) by RNA polymerase II followed by preliminary processing by a nuclear enzyme Drosha. The resulting hairpin-loop precursor miRNAs (pre-miRNAs) are exported to cytoplasm by the Exportin-5, where Dicer, TRBP, and Argonaute (Ago) proteins further processed to miRNA duplexes. The duplex miRNAs are associated with RISC. One strand of each duplex remains on the Ago protein in the RISC as the mature miRNA; the other strand is degraded. The mature miRNA is guided to its target 3′ UTR of mRNA for the translational repression or mRNA degradation.

machinery to block protein production [27]. Because of the partial base pairing nature to recognize its target mRNA, single miRNA, in theory, can target and regulate up to hundreds of different mRNAs. In early dates, the translational repression role of miRNAs was mainly emphasized [21, 42]. However, later studies also found that miRNAs can cause mRNA degradation through poly-A tail deadenylation [43, 44]. Recent study reported that mRNA degradation by a miRNA can be predominant and irreversible [45]. By those two mechanisms mediated by miRNAs, the production of proteins can be regulated at the post-transcript level. Therefore, aberrant regulations on miRNA expression may cause serious change of intracellular signal pathways as frequently observed in human cancers [46].

12.3 MicroRNAs as Tumor Suppressors

Like a protein-coding gene, a miRNA can act as a tumor suppressor when its loss of function is related to malignant transformation of a normal cell. The loss of function of a miRNA can be a result of genomic deletion, mutation, epigenetic silencing, and/or miRNA processing alterations (Table 12.1). Two miRNAs, *miR-15a* and *miR-16-1,* was discovered from the 30-kb deleted region between exons 2 and 5 of the *LEU2* gene from 13q14.2 region, the most reported chromosomal abnormality in chronic lymphocytic leukemia (CLL) with frequent hemizygous or homozygous deletion which occurs in >50% of cases [5, 47, 48]. The *miR-15a/16-1* cluster were proposed to play a tumor suppressor role, since the *miR-15a/miR-16-1* cluster was found to target anti-apoptotic protein BCL-2, which is widely overexpressed in CLL. Therefore, low levels of *miR-15a/miR-16-1* may flourish BCL-2 protein expression in a subset of

Table 12.1 Representative tumor suppressor miRNAs in human cancers. *Abbreviations*: CLL, chronic lymphocytic leukemia; AML, acute myeloid leukemia; TS, tumor suppressor

MiRNA	Expression in patients	Targets	Function
miR-15a *miR-16-1*	Downregulated in CLL	Bcl-2, Wt-1	Induces apoptosis and decrease tumorigenicity
let-7 family	Downregulated in lung and breast cancer	RAS, c-myc, HMGA2	Induces apoptosis
miR-29a *miR-29b* *miR-29c*	Downregulated in CLL, AML (11q23), lung and breast cancers, and cholangiocarcinoma	TCL-1, MCL1, DNMT3s	Induces apoptosis and decrease tumorigenicity
miR-34a *miR-34b* *miR34c*	Downregulated in pancreatic, colon, and breast cancers	CDK4, CDK6, cyclinE2, E2F3	Induces apoptosis

CLL patients carrying genomic deletion or, less frequently, mutations [5, 47, 48]. In myelodysplasia and therapy-related acute myeloid leukemia (AML), miR-29 family, *miR-29b-1/miR-29a,* was found in chromosome 7q32 which is a frequently deleted region in AML [49, 50]. The downregulated expression of miR-29 family members was also inversely correlated with upregulated oncogenic proteins including BCL-2 and MCL-1 [51], suggesting their tumor suppressor role in AML. The miR-29 family is also frequently downregulated in lung cancers and found to target DNA methyltransferase *(DNMT)3A* and *-3B* (de novo methyltransferases), which are two key enzymes involved in DNA methylation, that are frequently upregulated in lung cancer and associated with poor prognosis [52].

Let-7 family includes 12 miRNAs and all of them are believed to function as tumor suppressors during cancer development in lung, breast, urothelial, and cervical cancers as they locate at fragile site that involve in those cancers [53]. Reduced expression of *let-7* miRNAs were observed in lung cancers [54], and their target, Ras, was upregulated to contribute to lung carcinogenesis [55]. Interestingly, most of the tumor suppressor miRNAs (*miR-15-a/16-1, miR-29s*, and *let-7*) have more than one genomic location that expresses identical mature miRNAs. Multi-copy of the tumor suppressor miRNA could be conserved through evolution conserved mechanism to preserve function of an important miRNA if one allele is deleted or silenced.

12.4 MiRNAs as Oncogenes

In contrast to the tumor suppressor miRNAs, significantly upregulated miRNAs in human cancers are considered to possess oncogenic roles (Table 12.2). The first miRNA proposed to play an oncogenic function was *miR-155,* since its accumulation was observed in human B-cell lymphomas together with a host noncoding RNA named the B cell integration cluster (*BIC*) located in chromosome 21q23 [56]. Later, it was also shown that *miR-155* is highly expressed in many other human cancers including pediatric Burkitt lymphoma, Hodgkin disease, primary mediastinal non-Hodgkin lymphoma, CLL, AML, lung cancer, and breast cancer [57].

Table 12.2 Representative oncogenic miRNAs in human cancers. *Abbreviations*: CLL, chronic lymphocytic leukemia; AML, acute myeloid leukemia; DLBCL, diffuse large B cell lymphoma; BL, Burkitt lymphoma

MiRNA	Expression in patients	Targets	Function
miR-155	Upregulated in CLL, DLBCL, AML, BL, and lung and breast cancers	FOXO3a SOCS1 SHIP1	Induces lymphoproliferation, pre-B lymphoma/ leukemia in mice
miR-21	Upregulated in breast, colon, pancreas, lung, prostate, liver, and stomach cancer; AML(11q23); CLL; and glioblastoma	PTEN, PDCD4, TPM1, TIMP3	Inhibits apoptosis and enhances tumorigenicity
miR-17-92 cluster	Upregulated in lymphomas and in breast, liver, lung, colon, stomach, and pancreas cancers	E2F1 Bim PTEN	Co-operates with c-myc to induce lymphoma in mice Develops lymphoproliferative disorder

A role for this miRNA in early leukemogenesis was proven in a transgenic mouse model with a B cell–targeted overexpression of *miR-155*, which underwent a polyclonal preleukemic pre–B-cell proliferation followed by full-blown B-cell malignancy [58]. More recently, it was reported in lymphoma study with *miR-155* overexpressing mice that tumor cell survival can be addicted to the expression of *miR-155* [59]. This addiction theory can be a huge benefit to cancer therapies, since targeting one specific oncogenic miRNA may be sufficient to inhibit entire tumor progression [59]. The addiction of cancer cells to a certain miRNA expression was first reported in *miR-21*. Overexpression of *miR-21* in spleen of mice has shown to develop lethal B-cell lymphoma, and more importantly the malignancy was addicted to the expression level of *miR-21* [60]. *MiR-21* is the most frequently upregulated miRNA in almost every type of human cancers including both hematopoietic and solid tumors, such as AML, CLL, glioblastoma, and cancers of the pancreas, prostate, stomach, colon, lung, breast, and liver [61]. Transgenic mice that are overexpressing *miR-21* throughout whole body showed no tumor formation. When the *miR-21* transgenic mice were crossed to activated k-Ras mutant mice, *miR-21* over-expression increased tumor incidents in lung cancer by negatively regulated multiple tumor suppressors (Spry1, Spry2, Btg2 and Pdcd4) through activated Ras/Mek/Erk pathway [62] consistent with the previous report [63].

Single cistronic m*iR-17-92* cluster containing six miRNAs (*miR-17, miR-18a, miR-19a, miR-20a, miR-19b-1*, and *miR-92-1*) is located within 800 base pairs in the noncoding gene C13orf25 at 13q31.3 [64]. This region is frequently amplified in follicular lymphoma and diffuse large B cell lymphoma. In addition to being key players in lung development and in regulation of the immune and hematopoietic systems, members of the *miR-17-92* cluster are highly expressed in a variety of solid tumors and hematological malignancies, including cancers of the breast, colon, lung, pancreas, prostate, and stomach as well as lymphomas [64]. These miRNAs promote proliferation, inhibit apoptosis, induce tumor angiogenesis, and cooperate with *c-myc* to cause lymphoma in mice [65]. Interestingly, the *miR-17-92* cluster is transactivated by *c-myc*, an oncogene that is frequently activated in cancer [66]. The effects of this cluster's expression on cell cycle and proliferation are partly due to its regulation of E2F transcription factors, which are critical regulators of the cell cycle [67]. Conversely, both *E2F1* and *E2F3* were shown to activate the *miR-17-92* cluster, establishing a regulatory loop [68, 69]. These oncogenic miRNAs have been studied as promising therapeutic targets, since their downregulation or knock-down have been demonstrated to induce apoptosis, cell cycle arrest, and inhibition of invasion and metastasis of cancer cells.

12.5 MiRNAs and Epigenetics

Expression of miRNA is also regulated by promoter DNA hypermethylation. For example, *miR-127* expression can be silenced by promoter hypermethylation in bladder cancer cell lines and patients, and hypomethylating agents restored its expression level [70]. MiRNAs themselves can modulate DNA methylation in cancers by interfering with the DNA methylation machinery. Tumor suppressor miRNA *miR-29* family can induce global DNA hypomethylation by targeting *DNMT3A* and *-3B*, and rescue re-expression of tumor suppressor genes [52]. Interestingly, *miR-29*s are downregulated in lung cancer patients, and an inverse correlation was found between *miR-29s* and *DNMT3B* expression, suggesting that the downregulation of these miRNAs may contribute to increased DNMT3 levels, as well as hypermethylation and silencing of tumor suppressor genes in lung cancer [52].

12.6 MiRNA Signatures in Human Cancers

As the number of miRNA grew, the need of global profiling on all kinds of miRNA in human cancers has come to the fore. In 2003 by Carlo Croce laboratory, first customized microarray platform was built to carry out the genome-wide miRNA expression profiling in human cancer specimen. The customized microarray containing hundreds of human precursor and mature miRNA oligonucleotide probes was first applied to analyze a number of human B-cell chronic lymphocytic leukemia (CLL) samples, and found a unique miRNA expression pattern in the malignant tumors which was not observed in normal cells [47]. Followed by subsequent profiling on more CLL samples, it was also revealed that such miRNA expression patterns are strongly associated with both prognosis and progression of CLL in patients [48]. In the following years, massive profiling on various human cancer types were performed and revealed the "signatures" of miRNA expression pattern from most human cancer specimens with clear context-dependent manner by distinguishing their developmental origin from others [71]. For example, *miR-15a/16-1* clusters were frequently deleted and downregulated in CLL [47]. However, *miR-21* has been found to be upregulated in most types of human cancer. The different signatures of miRNA profiles also distinguish subtypes of cancer. For example, acute lymphoblastic leukemia (ALL) specimens can be further divided into three subtypes: (i) both t(9;22) *BCR/ABL*- and t(12;21) *TEL/AML1*-positives; (ii) T-cell ALL; and (iii) *MLL* gene rearrangement [71]. *MiR-221/222* cluster was found to be highly upregulated in non–small cell lung cancers and hepatocellular carcinomas [72].

Besides the hybridization-based microarray, quantitative real-time polymerase chain reaction (qRT-PCR) has been also used to profile miRNA signatures from human cancer samples. For example, Mestdagh et al. employed Megaplex reverse transcription format to use stem-loop-specific primers for qRT-PCR-based array after pre-amplified complementary DNA. The combination enabled to reduce input RNA amount down to 10 ng, which makes it a good choice for high-throughput miRNA signature profiling.

Massive qRT-PCR-based array was also used to pinpoint miRNA signatures at different regulation levels. Compared to one-way hybridization-based microarray which was concentrated to detect mature forms of miRNA, the qRT-PCR-based array seems more useful to track changes of miRNA expression on multiple levels. For example, O'Hara et al. used the qRT-PCR-based array to profile a set of primary effusion lymphomas (PELs) and found 68 miRNAs as PEL-specific signatures [73]. They observed that the 68 miRNA signatures are amplified at three levels of expression levels: gene alteration, transcription (pre-miRNA) and processing (mature miRNA). This qRT-PCR-based array uses a set of primers of interest, therefore it is relatively easy to customize the array set to specify for a narrow range of pathways or pathogens [74].

These findings inspired oncologists by an idea that the miRNA profiling may serve as a simple step diagnostic tool to screen potential cancer patients with specific type of cancer that the patient may suffer from [46].

12.7 Biomarker MiRNAs in Human Cancers

Early detection of cancerous phenotype is one of the key factors in cancer diagnosis that decide positive outcomes of cancer treatment options, since late detection often results poor prognosis due to metastasis to other organs. Global profiling of overall miRNAs is massive task to be performed on every patient samples in terms of time and cost. Narrow range of few miRNAs can be easily carried out in a short time period, which will greatly increase the rate of early detection. In this regard, solid knowledge of accurate and affordable biomarker on each type of human cancers is critical. MicroRNA signature has been proposed for this role. Discoveries of existence of miRNA signatures in various types of human cancer prompted many researchers to go in deep to pinpoint the most critical miRNAs even after excluding heterogeneous genetic and historic background of different specimens. If it is possibly doable, such miRNAs can be useful as biomarkers in determination of diagnosis and prognosis. For example, *miR-15/16* clusters were found to be frequently deleted and downregulated in CLL [47]. Comparison of miRNA profiles from 166 human bladder tumor samples to 11 normal bladder samples revealed 15 miRNAs that robustly deregulated in bladder cancer, and only 3 miRNAs (*miR-9, miR-182* and *miR-200b*) were found to be the miRNA signature associated with tumor aggressiveness. Those three miRNA signatures can be useful prognostic markers. Although obtaining miRNA signatures from vast profiling on a number of patient samples can be useful tools as diagnostic and prognostic markers, there are still huge variations on the profiling results among research groups. Early miRNA profiling study on lung adenocarcinoma samples revealed that high *miR-155* and low *let-7a-2* expression are strongly correlated with poor survival [75]. However, recent profiling on 372 TCGA data proposed eight miRNAs (*miR-31, miR-196b, miR-766, miR-519a-1, miR-375, miR-187, miR-331* and *miR-101-1*) as the best prognostic markers [76]. This case indicates that miRNA signatures can be affected by a number of factors, such as sample size, genetic background, detection method, and analysis method. Therefore, deeper breakdown on sample groups based on their genetic or historic background may need to be considered. For example, the miRNA/TP53 feedback regulatory circuitry was found in CLL patient samples with tight regulation mechanism [77]. *MiR-15a/16-1* cluster and *miR-34b/34c* cluster, dysregulated in CLL, were activated by TP53, and the *miR-34b/34c* cluster could directly target ZAP70 kinase. Interestingly, those miRNAs and the target genes in the miRNA-TP53 feedback circuitry were located at the regions that are frequently deleted in CLL (*miR-15a/miR-16-1* cluster at 13q, TP53 at 17p, and *miR-34b/miR-34c* cluster at 11q) [77]. Knowing genetic background should provide onco-clinicians to find a better treatment option for those patients. Correlating miRNA signatures with specific genetic background is also found to be useful for prognostic determination. For example, triple negative breast cancers (TNBC) were found to associate with four aberrantly regulated miRNAs (*miR-155, miR-493, miR-30e* and *miR-27a*) that were proposed as prognostic markers to distinguish TNBC patients from other breast cancer patients. These examples suggest that miRNA signatures may be useful to narrow down a specific subtype of cancers that will help to personalize an optimized treatment.

12.8 Circulating MiRNAs as Biomarkers of Human Cancers

Most aforementioned miRNA profiling was achieved on RNA samples directly collected from patients' cancer tissue obtained by biopsies. Alternatively, circulating RNAs, meaning RNA extracts from plasma and serum, was proposed as a non-invasive, simple, low-cost and rapid diagnostic tool for cancer detection [78]. First evidence of circulating miRNAs as diagnostic biomarkers was provided from miRNA profiling on serum RNA samples from diffuse large B-cell lymphoma (DLBCL) patients, which found three miRNA signatures (*miR-155, miR-210* and *miR-21*) that were highly elevated in DCBCL compared to normal sera [79]. Similar approach was used on exosome miRNA profiling. Taylor et al. isolated ovarian tumor-derived serum exosomes by using anti-EpCam antibody, and found 8 microRNA signatures (*miR-21, miR-141, miR-200a, miR-200c, miR-200b, miR-203, miR-205* and *miR-214*) as tumor-related biomarkers [80]. In other study, *miR-17-3p* and *miR-92* were found as miRNA signatures in serum RNAs from colorectal cancer (CRC) patients [81]. In hepatocellular carcinoma (HCC), *miR-21* was identified as the most enriched miRNA in plasma [82] agreed to the previous profiling result performed in HCC tissues miRNAs [83]. The authors proposed that the combinatory detection of circulating *miR-21* with serum α-feroprotein will improve an accuracy and specificity of biochemical markers for HCC. However, discrepancy of miRNA signatures between tissue and serum sample have been also reported. For example, breast cancer–derived serum miRNA signatures were only overlapped with 7 same miRNAs out of top 20 miRNAs that were significantly dysregulated in tissues and 16 out of top 20 miRNAs in serum were novel [84]. The limited overlaps of miRNA signatures between tissue and serum samples or discrepancy will bring about a questioning on a successful application of circulating miRNAs for diagnostic and prognostic tools. Couples of reason for these matters were offered, mostly based on methodological questions in regards to sampling methods, array methods and validation methods [85, 86]. However, in another view, it may be originated from the nature of heterogeneity of cancer origins and genetic backgrounds of each patient. Some miRNA signatures as prognostic predictors related to survival rate have been already suggested. *Mir-221* was shown to be useful to predict the development of lethal prostate cancers than indolent forms [87]. *MiR-150/miR-886-3p* signatures were proposed as prognostic predictors after surgery and chemotherapy in small cell lung cancer patients [88].

12.9 Single Polymorphism (SNP) in MiRNAs

In 2002, miR-*15a* and *miR-16-1* was discovered from the 30-kb deleted region between exons 2 and 5 of the *LEU2* gene in the 13q14.2 region [47]. In addition to the deletion, polymorphic genetic mutations of miRNA binding sites in 3′ UTR of target mRNAs were proposed to be causative variants for human cancers [89, 90]. When the regions harboring two microRNA-binding sites in the *KIT* gene from papillary thyroid carcinoma patients were sequenced, a polymorphism of G > A SNP (rs17084733) was found in the *KIT* 3′ UTR complementary sequences to the seed sequences of two miRNAs, *miR-221/222* [66]. In 5 out of 10 cases, this germ line mutation increased the association between *miR-221/222* and the target gene and was believed to contribute to the cancer progression. When over 100 human cancer samples and cancer cell lines were examined by sequence variation comparison, it was found that the SNP (G < A mutation) at 19 base pair of *Let-7* miRNA dramatically reduced the expression level of *Let-7* [91]. Global comparison of SNP frequencies between human cancer EST libraries and SNP databases has identified 12 miRNAs with SNPs that were the most aberrantly regulated in human cancers. In the other hand, SNP in miRNA sequences also can contribute to tumorigenesis. In CLL, it was found that the *miR-16-1* carries germ line mutation in high frequency [48]. Overall frequency of SNP in miRNA-binding site or SNP in miRNA sequences was predicted to be less than 1%, due to the size of miRNA and its binding site [90]. However, successful pinpointing such SNP in miRNA and its targets will provide a fine-tuned personalized miRNA signature for each cancer patient.

12.10 MiRNA and Cancer Stem Cells

Global miRNA profiling has revealed that miRNA functions in the regulation of cell development and the maintenance of pluripotency. In 2003, stem cell-specific miRNA signature was first described in mouse embryonic stem cells [92]. Involvement of miRNAs during cancer stem cell (CSC) formation and maintenance was also reported. For example, *let-7* expression was found to be reduced in breast cancer stem cells and its downregulation released the overexpression of *H-RAS* and *HMGA2* [93]. The upregulation of *H-RAS* and *HMGA2* increased stem cell-like self-renewal and reduced differentiation properties. Another miRNA, *miR-200b*, was also found to be a cancer-specific signature in breast cancers and epigenetically lost during CSC formation [94]. Mechanistically, the loss of *miR-200* upregulates its direct target gene expression, SUZ12, which epigenetically mediates polycomb-mediated repression of the E-cadherin gene [94]. Since CSCs have been thought to initiate tumor formation and induce cancer recurrences, CSC-related miRNA signatures can serve as diagnostic/prognostic markers and therapeutic targets.

12.11 Perspectives

Ever since the first discovery of cancer-involved *miR-15a/16-1* cluster in CLL, the small molecules opened a new big world in cancer research. Roles of miRNAs in human cancers have been

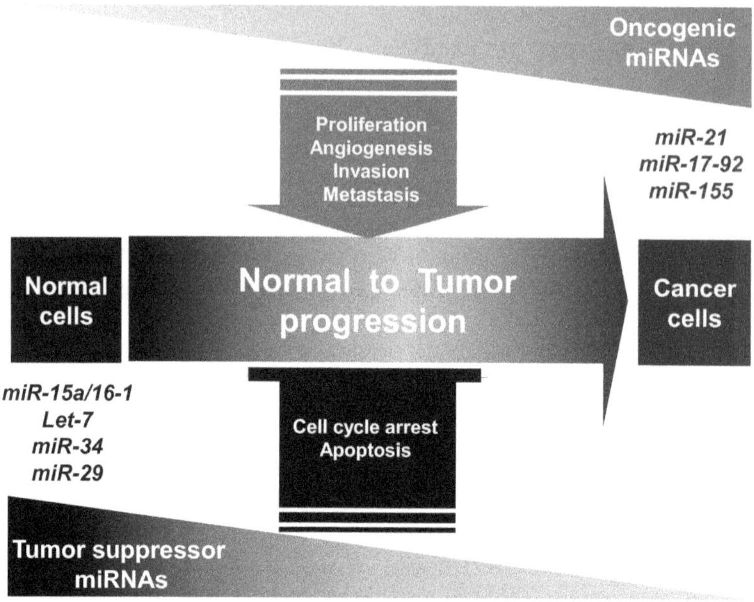

Figure 12.2 Functional regulation of human cancer cells by miRNAs during malignant transformation, progression, invasion and metastasis. Tumor suppressor miRNAs are downregulated to release the burst of oncogenic protein expression, while oncogenic miRNAs are upregulated to suppress tumor suppressor proteins. Orchestration of these deregulated miRNAs promotes inhibition of apoptosis, and tumor survivals.

shown to involve almost all aspects of cancer biology, including proliferation, apoptosis, invasion/metastasis, angiogenesis, and even CSC formation and maintenance (Fig. 12.2). More importantly, the expression patterns of certain miRNAs are closely correlated with tumorigenesis and progression. Massive global profiling of miRNA expressions in a variety of human cancer patient samples found unique signature of miRNAs that are deregulated in the cancers. Identification of definitive miRNA signatures can serve as diagnostic and prognostic tools, and therapeutic targets. However, most examples of miRNA signatures in cancers obtained from global miRNA profiling on massive patients samples compared to normal to every patient reveal the weak point in terms of variation and discrepancy among different studies, which is still remained to resolve before routine application of miRNA signatures to bench-side screening. One way to compromise the challenging questions is to increase the profiling data size by analyzing a large number of samples in all ethnic, sex and age groups with compared to without cancers. Another way is to more personalize such profiling effort into individual cases. Previous pathological and/or treatment history of individual cancer patients often provides quite useful information to define risk factors or predictors. In addition, profiling data from long-term observation from human groups will fine-tune the selection of miRNA signatures. For example, comparison of the miRNA profiles before and after a cancer occurrence, and every stages of cancer development will pin-point such deregulated miRNAs during the whole events. MiRNA-based cancer therapies are under many trials, and it is too early to expect to replace the conventional treatment options by themselves. However, upcoming achievements and progress about miRNA related to human cancers will provide a new option for diagnostic/prognostic tools and therapeutic strategies.

References

1. Cech, T. R., and Steitz, J. A. The noncoding RNA revolution-trashing old rules to forge new ones. *Cell* 2014, **157**, 77–94.
2. Gibb, E. A., Brown, C. J., and Lam, W. L. The functional role of long non-coding RNA in human carcinomas. *Mol. Cancer.* 2011, **10**, 38-4598-10-38.
3. Brannan, C. I., Dees, E. C., Ingram, R. S., and Tilghman, S. M. The product of the H19 gene may function as an RNA. *Mol. Cell. Biol.* 1990, **10**, 28–36.
4. Lee, R. C., Feinbaum, R. L., and Ambros, V. The C. elegans heterochronic gene lin-4 encodes small RNAs with antisense complementarity to lin-14. *Cell* 1993, **75**, 843–854.
5. Calin, G. A., Dumitru, C. D., Shimizu, M., Bichi, R., Zupo, S., Noch, E., Aldler, H., Rattan, S., Keating, M., Rai, K., Rassenti, L., Kipps, T., Negrini, M., Bullrich, F., and Croce, C. M. Frequent deletions and down-regulation of micro- RNA genes miR15 and miR16 at 13q14 in chronic lymphocytic leukemia. *Proc. Natl. Acad. Sci. U. S. A.* 2002, **99**, 15524–15529.

6. Barsyte-Lovejoy, D., Lau, S. K., Boutros, P. C., Khosravi, F., Jurisica, I., Andrulis, I. L., Tsao, M. S., and Penn, L. Z. The c-Myc oncogene directly induces the H19 noncoding RNA by allele-specific binding to potentiate tumorigenesis. *Cancer Res.* 2006, **66**, 5330–5337.
7. Dugimont, T., Montpellier, C., Adriaenssens, E., Lottin, S., Dumont, L., Iotsova, V., Lagrou, C., Stehelin, D., Coll, J., and Curgy, J. J. The H19 TATA-less promoter is efficiently repressed by wild-type tumor suppressor gene product p53. *Oncogene* 1998, **16**, 2395–2401.
8. Gupta, R. A., Shah, N., Wang, K. C., Kim, J., Horlings, H. M., Wong, D. J., Tsai, M. C., Hung, T., Argani, P., Rinn, J. L., Wang, Y., Brzoska, P., Kong, B., Li, R., West, R. B., van de Vijver, M. J., Sukumar, S., and Chang, H. Y. Long non-coding RNA HOTAIR reprograms chromatin state to promote cancer metastasis. *Nature* 2010, **464**, 1071–1076.
9. Brown, C. J., Ballabio, A., Rupert, J. L., Lafreniere, R. G., Grompe, M., Tonlorenzi, R., and Willard, H. F. A gene from the region of the human X inactivation centre is expressed exclusively from the inactive X chromosome. *Nature* 1991, **349**, 38–44.
10. Froberg, J. E., Yang, L., and Lee, J. T. Guided by RNAs: X-inactivation as a model for lncRNA function. *J. Mol. Biol.* 2013, **425**, 3698–3706.
11. Ji, P., Diederichs, S., Wang, W., Boing, S., Metzger, R., Schneider, P. M., Tidow, N., Brandt, B., Buerger, H., Bulk, E., Thomas, M., Berdel, W. E., Serve, H., and Muller-Tidow, C. MALAT-1, a novel noncoding RNA, and thymosin beta4 predict metastasis and survival in early-stage non-small cell lung cancer. *Oncogene* 2003, **22**, 8031–8041.
12. Pang, E. J., Yang, R., Fu, X. B., and Liu, Y. F. Overexpression of long non-coding RNA MALAT1 is correlated with clinical progression and unfavorable prognosis in pancreatic cancer. *Tumour Biol.* 2014.
13. Wang, X., Li, M., Wang, Z., Han, S., Tang, X., Ge, Y., Zhou, L., Zhou, C., Yuan, Q., and Yang, M. Silencing of long noncoding RNA MALAT1 by miR-101 and miR-217inhibits proliferation, migration and invasion of esophageal squamous cell carcinoma cells. *J. Biol. Chem.* 2014.
14. Zhang, H. M., Yang, F. Q., Chen, S. J., Che, J., and Zheng, J. H. Upregulation of long non-coding RNA MALAT1 correlates with tumor progression and poor prognosis in clear cell renal cell carcinoma. *Tumour Biol.* 2014.
15. Zheng, H. T., Shi, D. B., Wang, Y. W., Li, X. X., Xu, Y., Tripathi, P., Gu, W. L., Cai, G. X., and Cai, S. J. High expression of lncRNA MALAT1 suggests a biomarker of poor prognosis in colorectal cancer. *Int. J. Clin. Exp. Pathol.* 2014, **7**, 3174–3181.

16. Zhang, X., Zhou, Y., Mehta, K. R., Danila, D. C., Scolavino, S., Johnson, S. R., and Klibanski, A. A pituitary-derived MEG3 isoform functions as a growth suppressor in tumor cells. *J. Clin. Endocrinol. Metab.* 2003, **88**, 5119–5126.
17. Sun, M., Xia, R., Jin, F., Xu, T., Liu, Z., De, W., and Liu, X. Downregulated long noncoding RNA MEG3 is associated with poor prognosis and promotes cell proliferation in gastric cancer. *Tumour Biol.* 2014, **35**, 1065–1073.
18. Ying, L., Huang, Y., Chen, H., Wang, Y., Xia, L., Chen, Y., Liu, Y., and Qiu, F. Downregulated MEG3 activates autophagy and increases cell proliferation in bladder cancer. *Mol. Biosyst.* 2013, **9**, 407–411.
19. Anwar, S. L., Krech, T., Hasemeier, B., Schipper, E., Schweitzer, N., Vogel, A., Kreipe, H., and Lehmann, U. Loss of imprinting and allelic switching at the DLK1-MEG3 locus in human hepatocellular carcinoma. *PLoS One* 2012, **7**, e49462.
20. Wang, P., Ren, Z., and Sun, P. Overexpression of the long non-coding RNA MEG3 impairs in vitro glioma cell proliferation. *J. Cell. Biochem.* 2012, **113**, 1868–1874.
21. Wightman, B., Ha, I., and Ruvkun, G. Posttranscriptional regulation of the heterochronic gene lin-14 by lin-4 mediates temporal pattern formation in C. elegans. *Cell* 1993, *75*, 855–862.
22. Hamilton, A. J., and Baulcombe, D. C. A species of small antisense RNA in posttranscriptional gene silencing in plants. *Science* 1999, **286**, 950–952.
23. Elbashir, S. M., Harborth, J., Lendeckel, W., Yalcin, A., Weber, K., and Tuschl, T. Duplexes of 21-nucleotide RNAs mediate RNA interference in cultured mammalian cells. *Nature* 2001, **411**, 494–498.
24. Lee, R. C., and Ambros, V. An extensive class of small RNAs in Caenorhabditis elegans. *Science* 2001, **294**, 862–864.
25. Lau, N. C., Lim, L. P., Weinstein, E. G., and Bartel, D. P. An abundant class of tiny RNAs with probable regulatory roles in Caenorhabditis elegans. *Science* 2001, **294**, 858–862.
26. Lagos-Quintana, M., Rauhut, R., Lendeckel, W., and Tuschl, T. Identification of novel genes coding for small expressed RNAs. *Science* 2001, **294**, 853–858.
27. Bartel, D. P. MicroRNAs: target recognition and regulatory functions. *Cell* 2009, **136**, 215–233.
28. Hatfield, S. D., Shcherbata, H. R., Fischer, K. A., Nakahara, K., Carthew, R. W., and Ruohola-Baker, H. Stem cell division is regulated by the microRNA pathway. *Nature* 2005, **435**, 974–978.

29. Kawasaki, H., and Taira, K. Hes1 is a target of microRNA-23 during retinoic-acid-induced neuronal differentiation of NT2 cells. *Nature* 2003, **423**, 838–842.

30. Chen, C. Z., Li, L., Lodish, H. F., and Bartel, D. P. MicroRNAs modulate hematopoietic lineage differentiation. *Science* 2004, **303**, 83–86.

31. Boehm, M., and Slack, F. J. MicroRNA control of lifespan and metabolism. *Cell. Cycle* 2006, **5**, 837–840.

32. Poy, M. N., Eliasson, L., Krutzfeldt, J., Kuwajima, S., Ma, X., Macdonald, P. E., Pfeffer, S., Tuschl, T., Rajewsky, N., Rorsman, P., and Stoffel, M. A pancreatic islet-specific microRNA regulates insulin secretion. *Nature* 2004, **432**, 226–230.

33. van Rooij, E., Sutherland, L. B., Liu, N., Williams, A. H., McAnally, J., Gerard, R. D., Richardson, J. A., and Olson, E. N. A signature pattern of stress-responsive microRNAs that can evoke cardiac hypertrophy and heart failure. *Proc. Natl. Acad. Sci. U. S. A.* 2006, **103**, 18255–18260.

34. Caudy, A. A., Myers, M., Hannon, G. J., and Hammond, S. M. Fragile X-related protein and VIG associate with the RNA interference machinery. *Genes Dev.* 2002, **16**, 2491–2496.

35. Bilen, J., Liu, N., Burnett, B. G., Pittman, R. N., and Bonini, N. M. MicroRNA pathways modulate polyglutamine-induced neurodegeneration. *Mol. Cell* 2006, **24**, 157–163.

36. Calin, G. A., and Croce, C. M. MicroRNA signatures in human cancers. *Nat. Rev. Cancer.* 2006, **6**, 857–866.

37. Lee, Y., Kim, M., Han, J., Yeom, K. H., Lee, S., Baek, S. H., and Kim, V. N. MicroRNA genes are transcribed by RNA polymerase II. *EMBO J.* 2004, **23**, 4051–4060.

38. Lee, Y., Ahn, C., Han, J., Choi, H., Kim, J., Yim, J., Lee, J., Provost, P., Radmark, O., Kim, S., and Kim, V. N. The nuclear RNase III Drosha initiates microRNA processing. *Nature* 2003, **425**, 415–419.

39. Yi, R., Qin, Y., Macara, I. G., and Cullen, B. R. Exportin-5 mediates the nuclear export of pre-microRNAs and short hairpin RNAs. *Genes Dev.* 2003, **17**, 3011–3016.

40. Hutvagner, G., McLachlan, J., Pasquinelli, A. E., Balint, E., Tuschl, T., and Zamore, P. D. A cellular function for the RNA-interference enzyme Dicer in the maturation of the let-7 small temporal RNA. *Science* 2001, **293**, 834–838.

41. Ketting, R. F., Fischer, S. E., Bernstein, E., Sijen, T., Hannon, G. J., and Plasterk, R. H. Dicer functions in RNA interference and in synthesis of

small RNA involved in developmental timing in C. elegans. *Genes Dev.* 2001, **15**, 2654–2659.

42. Olsen, P. H., and Ambros, V. The lin-4 regulatory RNA controls developmental timing in Caenorhabditis elegans by blocking LIN-14 protein synthesis after the initiation of translation. *Dev. Biol.* 1999, **216**, 671–680.
43. Bagga, S., Bracht, J., Hunter, S., Massirer, K., Holtz, J., Eachus, R., and Pasquinelli, A. E. Regulation by let-7 and lin-4 miRNAs results in target mRNA degradation. *Cell* 2005, **122**, 553–563.
44. Giraldez, A. J., Mishima, Y., Rihel, J., Grocock, R. J., Van Dongen, S., Inoue, K., Enright, A. J., and Schier, A. F. Zebrafish MiR-430 promotes deadenylation and clearance of maternal mRNAs. *Science* 2006, **312**, 75–79.
45. Eichhorn, S. W., Guo, H., McGeary, S. E., Rodriguez-Mias, R. A., Shin, C., Baek, D., Hsu, S. H., Ghoshal, K., Villen, J., and Bartel, D. P. mRNA destabilization is the dominant effect of mammalian microRNAs by the time substantial repression ensues. *Mol. Cell* 2014, **56**, 104–115.
46. Croce, C. M. Causes and consequences of microRNA dysregulation in cancer. *Nat. Rev. Genet.* 2009, **10**, 704–714.
47. Calin, G. A., Liu, C. G., Sevignani, C., Ferracin, M., Felli, N., Dumitru, C. D., Shimizu, M., Cimmino, A., Zupo, S., Dono, M., Dell'Aquila, M. L., Alder, H., Rassenti, L., Kipps, T. J., Bullrich, F., Negrini, M., and Croce, C. M. MicroRNA profiling reveals distinct signatures in B cell chronic lymphocytic leukemias. *Proc. Natl. Acad. Sci. U. S. A.* 2004, **101**, 11755–11760.
48. Calin, G. A., Ferracin, M., Cimmino, A., Di Leva, G., Shimizu, M., Wojcik, S. E., Iorio, M. V., Visone, R., Sever, N. I., Fabbri, M., Iuliano, R., Palumbo, T., Pichiorri, F., Roldo, C., Garzon, R., Sevignani, C., Rassenti, L., Alder, H., Volinia, S., Liu, C. G., Kipps, T. J., Negrini, M., and Croce, C. M. A MicroRNA signature associated with prognosis and progression in chronic lymphocytic leukemia. *N. Engl. J. Med.* 2005, **353**, 1793–1801.
49. Garzon, R., Garofalo, M., Martelli, M. P., Briesewitz, R., Wang, L., Fernandez-Cymering, C., Volinia, S., Liu, C. G., Schnittger, S., Haferlach, T., Liso, A., Diverio, D., Mancini, M., Meloni, G., Foa, R., Martelli, M. F., Mecucci, C., Croce, C. M., and Falini, B. Distinctive microRNA signature of acute myeloid leukemia bearing cytoplasmic mutated nucleophosmin. *Proc. Natl. Acad. Sci. U. S. A.* 2008, **105**, 3945–3950.
50. Garzon, R., Heaphy, C. E., Havelange, V., Fabbri, M., Volinia, S., Tsao, T., Zanesi, N., Kornblau, S. M., Marcucci, G., Calin, G. A., Andreeff, M., and

Croce, C. M. MicroRNA 29b functions in acute myeloid leukemia. *Blood* 2009, **114**, 5331–5341.

51. Xu, L., Xu, Y., Jing, Z., Wang, X., Zha, X., Zeng, C., Chen, S., Yang, L., Luo, G., Li, B., and Li, Y. Altered expression pattern of miR-29a, miR-29b and the target genes in myeloid leukemia. *Exp. Hematol. Oncol.* 2014, **3**, 17-3619-3-17. eCollection 2014.

52. Fabbri, M., Garzon, R., Cimmino, A., Liu, Z., Zanesi, N., Callegari, E., Liu, S., Alder, H., Costinean, S., Fernandez-Cymering, C., Volinia, S., Guler, G., Morrison, C. D., Chan, K. K., Marcucci, G., Calin, G. A., Huebner, K., and Croce, C. M. MicroRNA-29 family reverts aberrant methylation in lung cancer by targeting DNA methyltransferases 3A and 3B. *Proc. Natl. Acad. Sci. U. S. A.* 2007, **104**, 15805–15810.

53. Calin, G. A., Sevignani, C., Dumitru, C. D., Hyslop, T., Noch, E., Yendamuri, S., Shimizu, M., Rattan, S., Bullrich, F., Negrini, M., and Croce, C. M. Human microRNA genes are frequently located at fragile sites and genomic regions involved in cancers. *Proc. Natl. Acad. Sci. U. S. A.* 2004, **101**, 2999–3004.

54. Takamizawa, J., Konishi, H., Yanagisawa, K., Tomida, S., Osada, H., Endoh, H., Harano, T., Yatabe, Y., Nagino, M., Nimura, Y., Mitsudomi, T., and Takahashi, T. Reduced expression of the let-7 microRNAs in human lung cancers in association with shortened postoperative survival. *Cancer Res.* 2004, **64**, 3753–3756.

55. Johnson, S. M., Grosshans, H., Shingara, J., Byrom, M., Jarvis, R., Cheng, A., Labourier, E., Reinert, K. L., Brown, D., and Slack, F. J. RAS is regulated by the let-7 microRNA family. *Cell* 2005, **120**, 635–647.

56. Eis, P. S., Tam, W., Sun, L., Chadburn, A., Li, Z., Gomez, M. F., Lund, E., and Dahlberg, J. E. Accumulation of miR-155 and BIC RNA in human B cell lymphomas. *Proc. Natl. Acad. Sci. U. S. A.* 2005, **102**, 3627–3632.

57. Jurkovicova, D., Magyerkova, M., Kulcsar, L., Krivjanska, M., Krivjansky, V., Gibadulinova, A., Oveckova, I., and Chovanec, M. miR-155 as a diagnostic and prognostic marker in hematological and solid malignancies. *Neoplasma* 2014, **61**, 241–251.

58. Costinean, S., Zanesi, N., Pekarsky, Y., Tili, E., Volinia, S., Heerema, N., and Croce, C. M. Pre-B cell proliferation and lymphoblastic leukemia/high-grade lymphoma in E(mu)-miR155 transgenic mice. *Proc. Natl. Acad. Sci. U. S. A.* 2006, **103**, 7024–7029.

59. Cheng, C. J., Bahal, R., Babar, I. A., Pincus, Z., Barrera, F., Liu, C., Svoronos, A., Braddock, D. T., Glazer, P. M., Engelman, D. M., Saltzman, W. M., and

Slack, F. J. MicroRNA silencing for cancer therapy targeted to the tumour microenvironment. *Nature* 2014.

60. Medina, P. P., Nolde, M., and Slack, F. J. OncomiR addiction in an in vivo model of microRNA-21-induced pre-B-cell lymphoma. *Nature* 2010, **467**, 86–90.
61. Krichevsky, A. M., and Gabriely, G. miR-21: a small multi-faceted RNA. *J. Cell. Mol. Med.* 2009, **13**, 39–53.
62. Hatley, M. E., Patrick, D. M., Garcia, M. R., Richardson, J. A., Bassel-Duby, R., van Rooij, E., and Olson, E. N. Modulation of K-Ras-dependent lung tumorigenesis by MicroRNA-21. *Cancer. Cell.* 2010, **18**, 282–293.
63. Thum, T., Gross, C., Fiedler, J., Fischer, T., Kissler, S., Bussen, M., Galuppo, P., Just, S., Rottbauer, W., Frantz, S., Castoldi, M., Soutschek, J., Koteliansky, V., Rosenwald, A., Basson, M. A., Licht, J. D., Pena, J. T., Rouhanifard, S. H., Muckenthaler, M. U., Tuschl, T., Martin, G. R., Bauersachs, J., and Engelhardt, S. MicroRNA-21 contributes to myocardial disease by stimulating MAP kinase signalling in fibroblasts. *Nature* 2008, **456**, 980–984.
64. Olive, V., Li, Q., and He, L. Mir-17-92: a polycistronic oncomir with pleiotropic functions. *Immunol. Rev.* 2013, **253**, 158–166.
65. Xiao, C., Srinivasan, L., Calado, D. P., Patterson, H. C., Zhang, B., Wang, J., Henderson, J. M., Kutok, J. L., and Rajewsky, K. Lymphoproliferative disease and autoimmunity in mice with increased miR-17-92 expression in lymphocytes. *Nat. Immunol.* 2008, **9**, 405–414.
66. He, L., Thomson, J. M., Hemann, M. T., Hernando-Monge, E., Mu, D., Goodson, S., Powers, S., Cordon-Cardo, C., Lowe, S. W., Hannon, G. J., and Hammond, S. M. A microRNA polycistron as a potential human oncogene. *Nature* 2005, **435**, 828–833.
67. O'Donnell, K. A., Wentzel, E. A., Zeller, K. I., Dang, C. V., and Mendell, J. T. c-Myc-regulated microRNAs modulate E2F1 expression. *Nature* 2005, **435**, 839–843.
68. Woods, K., Thomson, J. M., and Hammond, S. M. Direct regulation of an oncogenic micro-RNA cluster by E2F transcription factors. *J. Biol. Chem.* 2007, **282**, 2130–2134.
69. Sylvestre, Y., De Guire, V., Querido, E., Mukhopadhyay, U. K., Bourdeau, V., Major, F., Ferbeyre, G., and Chartrand, P. An E2F/miR-20a autoregulatory feedback loop. *J. Biol. Chem.* 2007, **282**, 2135–2143.
70. Saito, Y., Liang, G., Egger, G., Friedman, J. M., Chuang, J. C., Coetzee, G. A., and Jones, P. A. Specific activation of microRNA-127 with

downregulation of the proto-oncogene BCL6 by chromatin-modifying drugs in human cancer cells. *Cancer. Cell.* 2006, **9**, 435–443.

71. Lu, J., Getz, G., Miska, E. A., Alvarez-Saavedra, E., Lamb, J., Peck, D., Sweet-Cordero, A., Ebert, B. L., Mak, R. H., Ferrando, A. A., Downing, J. R., Jacks, T., Horvitz, H. R., and Golub, T. R. MicroRNA expression profiles classify human cancers. *Nature* 2005, **435**, 834–838.

72. Garofalo, M., Di Leva, G., Romano, G., Nuovo, G., Suh, S. S., Ngankeu, A., Taccioli, C., Pichiorri, F., Alder, H., Secchiero, P., Gasparini, P., Gonelli, A., Costinean, S., Acunzo, M., Condorelli, G., and Croce, C. M. miR-221&222 regulate TRAIL resistance and enhance tumorigenicity through PTEN and TIMP3 downregulation. *Cancer. Cell.* 2009, **16**, 498–509.

73. O'Hara, A. J., Vahrson, W., and Dittmer, D. P. Gene alteration and precursor and mature microRNA transcription changes contribute to the miRNA signature of primary effusion lymphoma. *Blood* 2008, **111**, 2347–2353.

74. Chugh, P., Tamburro, K., and Dittmer, D. P. Profiling of pre-micro RNAs and microRNAs using quantitative real-time PCR (qPCR) arrays. *J. Vis. Exp.* 2010, (46). pii: 2210. doi, 10.3791/2210.

75. Yanaihara, N., Caplen, N., Bowman, E., Seike, M., Kumamoto, K., Yi, M., Stephens, R. M., Okamoto, A., Yokota, J., Tanaka, T., Calin, G. A., Liu, C. G., Croce, C. M., and Harris, C. C. Unique microRNA molecular profiles in lung cancer diagnosis and prognosis. *Cancer. Cell.* 2006, **9**, 189–198.

76. Li, X., Shi, Y., Yin, Z., Xue, X., and Zhou, B. An eight-miRNA signature as a potential biomarker for predicting survival in lung adenocarcinoma. *J. Transl. Med.* 2014, **12**, 159-5876-12-159.

77. Fabbri, M., Bottoni, A., Shimizu, M., Spizzo, R., Nicoloso, M. S., Rossi, S., Barbarotto, E., Cimmino, A., Adair, B., Wojcik, S. E., Valeri, N., Calore, F., Sampath, D., Fanini, F., Vannini, I., Musuraca, G., Dell'Aquila, M., Alder, H., Davuluri, R. V., Rassenti, L. Z., Negrini, M., Nakamura, T., Amadori, D., Kay, N. E., Rai, K. R., Keating, M. J., Kipps, T. J., Calin, G. A., and Croce, C. M. Association of a microRNA/TP53 feedback circuitry with pathogenesis and outcome of B-cell chronic lymphocytic leukemia. *JAMA* 2011, **305**, 59–67.

78. Tsang, J. C., and Lo, Y. M. Circulating nucleic acids in plasma/serum. *Pathology* 2007, **39**, 197–207.

79. Lawrie, C. H., Gal, S., Dunlop, H. M., Pushkaran, B., Liggins, A. P., Pulford, K., Banham, A. H., Pezzella, F., Boultwood, J., Wainscoat, J. S., Hatton, C. S., and Harris, A. L. Detection of elevated levels of tumour-associated

microRNAs in serum of patients with diffuse large B-cell lymphoma. *Br. J. Haematol.* 2008, **141**, 672–675.

80. Taylor, D. D., and Gercel-Taylor, C. MicroRNA signatures of tumor-derived exosomes as diagnostic biomarkers of ovarian cancer. *Gynecol. Oncol.* 2008, **110**, 13–21.

81. Ng, E. K., Chong, W. W., Jin, H., Lam, E. K., Shin, V. Y., Yu, J., Poon, T. C., Ng, S. S., and Sung, J. J. Differential expression of microRNAs in plasma of patients with colorectal cancer: a potential marker for colorectal cancer screening. *Gut* 2009, **58**, 1375–1381.

82. Tomimaru, Y., Eguchi, H., Nagano, H., Wada, H., Kobayashi, S., Marubashi, S., Tanemura, M., Tomokuni, A., Takemasa, I., Umeshita, K., Kanto, T., Doki, Y., and Mori, M. Circulating microRNA-21 as a novel biomarker for hepatocellular carcinoma. *J. Hepatol.* 2012, **56**, 167–175.

83. Meng, F., Henson, R., Wehbe-Janek, H., Ghoshal, K., Jacob, S. T., and Patel, T. MicroRNA-21 regulates expression of the PTEN tumor suppressor gene in human hepatocellular cancer. *Gastroenterology* 2007, **133**, 647–658.

84. Chan, M., Liaw, C. S., Ji, S. M., Tan, H. H., Wong, C. Y., Thike, A. A., Tan, P. H., Ho, G. H., and Lee, A. S. Identification of circulating microRNA signatures for breast cancer detection. *Clin. Cancer Res.* 2013, **19**, 4477–4487.

85. Pritchard, C. C., Cheng, H. H., and Tewari, M. MicroRNA profiling: approaches and considerations. *Nat. Rev. Genet.* 2012, **13**, 358–369.

86. Moldovan, L., Batte, K. E., Trgovcich, J., Wisler, J., Marsh, C. B., and Piper, M. Methodological challenges in utilizing miRNAs as circulating biomarkers. *J. Cell. Mol. Med.* 2014, **18**, 371–390.

87. Kneitz, B., Krebs, M., Kalogirou, C., Schubert, M., Joniau, S., van Poppel, H., Lerut, E., Kneitz, S., Scholz, C. J., Strobel, P., Gessler, M., Riedmiller, H., and Spahn, M. Survival in patients with high-risk prostate cancer is predicted by miR-221, which regulates proliferation, apoptosis, and invasion of prostate cancer cells by inhibiting IRF2 and SOCS3. *Cancer Res.* 2014, **74**, 2591–2603.

88. Bi, N., Cao, J., Song, Y., Shen, J., Liu, W., Fan, J., He, J., Shi, Y., Zhang, X., Lu, N., Zhan, Q., and Wang, L. A microRNA signature predicts survival in early stage small-cell lung cancer treated with surgery and adjuvant chemotherapy. *PLoS One* 2014, **9**, e91388.

89. Chen, K., and Rajewsky, N. Natural selection on human microRNA binding sites inferred from SNP data. *Nat. Genet.* 2006, **38**, 1452–1456.

90. Saunders, M. A., Liang, H., and Li, W. H. Human polymorphism at microRNAs and microRNA target sites. *Proc. Natl. Acad. Sci. U. S. A.* 2007, **104**, 3300–3305.

91. Wu, M., Jolicoeur, N., Li, Z., Zhang, L., Fortin, Y., L'Abbe, D., Yu, Z., and Shen, S. H. Genetic variations of microRNAs in human cancer and their effects on the expression of miRNAs. *Carcinogenesis* 2008, **29**, 1710–1716.

92. Houbaviy, H. B., Murray, M. F., and Sharp, P. A. Embryonic stem cell-specific microRNAs. *Dev. Cell.* 2003, **5**, 351–358.

93. Yu, F., Yao, H., Zhu, P., Zhang, X., Pan, Q., Gong, C., Huang, Y., Hu, X., Su, F., Lieberman, J., and Song, E. Let-7 Regulates Self Renewal and Tumorigenicity of Breast Cancer Cells. *Cell* 2007, **131**, 1109–1123.

94. Iliopoulos, D., Lindahl-Allen, M., Polytarchou, C., Hirsch, H. A., Tsichlis, P. N., and Struhl, K. Loss of miR-200 inhibition of Suz12 leads to polycomb-mediated repression required for the formation and maintenance of cancer stem cells. *Mol. Cell* 2010, **39**, 761–772.

Chapter 13

Pharmacogenomics of Tamoxifen

Hitoshi Zembutsu
Project for Development of Liquid Biopsy Diagnosis, Cancer Precision Medicine Center, Japanese Foundation for Cancer Research, 3-8-31 Ariake, Koto-ku, Tokyo 135-8550, Japan
hitoshi.zembutsu@jfcr.or.jp

Tamoxifen has mainly been used to treat or prevent the recurrence of estrogen receptor (ER)–positive breast cancers in patients. CYP2D6 is a key enzyme responsible for generating the potent active metabolite of tamoxifen, "endoxifen." Many studies report that decreased or null-function alleles of *CYP2D6* were associated with poor clinical outcome of breast cancer patients treated with tamoxifen. However, there are some discrepant reports questioning the association between *CYP2D6* genotype and tamoxifen efficacy. Genotype-guided dose adjustment study of tamoxifen provides the evidence that dose adjustment is useful for the patients carrying reduced or null allele of *CYP2D6* to maintain the effective endoxifen level. Furthermore, this chapter describes critical issues in pharmacogenomics studies as well as summarizes the results of the association of *CYP2D6* genotype with tamoxifen efficacy.

Cancer Genetics and Genomics for Personalized Medicine
Edited by Il-Jin Kim

ISBN 978-981-4669-87-0 (Hardcover), 978-1-315-36476-6 (eBook)
www.panstanford.com

13.1 What Is Tamoxifen?

Tamoxifen is an antagonist of the estrogen receptor in breast tissue via its active metabolite, hydroxytamoxifen or endoxifen. Tamoxifen, which has been clinically used for more than 50 years, is proven to be effective in the adjuvant treatment and metastatic tumor of breast cancer. Tamoxifen is suggested to be used in the treatment of patients with pre- and postmenopausal estrogen receptor (ER)–positive breast cancer. Adjuvant tamoxifen therapy is one of the major endocrine treatment options, especially for women who have ovarian estrogenic activity that cannot be regulated by aromatase inhibitors. Five-year or 10-year adjuvant tamoxifen treatment effectively reduces recurrence of ER-positive tumors [1, 2]. Moreover, tamoxifen is approved for the chemoprevention of breast cancer for women at high risk of developing the disease. Tamoxifen may cause relatively mild adverse effects compared to chemotherapy, but is occasionally (or often) severe requiring to discontinue the treatment. "Hot flash" is the most common adverse event observed in patients treated with tamoxifen.

13.2 Why Is Pharmacogenomics of Tamoxifen Important?

According to the result of ATLAS (Adjuvant Tamoxifen Longer Against the Shorter), the risk of recurrence during years from 5 to 14 was greater than 20% in the patients treated with adjuvant tamoxifen therapy [2]. There are many studies with supportive results in the tamoxifen efficacy; however, the mechanisms underlying the resistance to tamoxifen in a subset of the patients are not fully identified. Tamoxifen is on some level considered as a prodrug because it requires metabolic activation to induce pharmacological activity. Endoxifen (4-hydroxy-*N*-desmethyltamoxifen) and 4-hydroxytamoxifen, the two major metabolites of tamoxifen, are reported to be active therapeutic moieties [3]. These two metabolites have 100-fold greater affinity to ER and 30- to 100-fold greater potency in inhibiting ER-positive cell growth compared to another prodrug, tamoxifen [4, 5]. It had been suggested that the differences in the amount of these active metabolites could affect the interindividual variability in the response to tamoxifen. Cytochrome P450 2D6 (CYP2D6) is an important enzyme in generating 4-hydroxytamoxifen and endoxifen. Genetic polymorphisms of *CYP2D6*, including alleles that affect the function and/or amount of the gene product, have been reported (http://www.cypalleles.ki.se/cyp2d6.htm). There are three groups based on the CYP2D6 phenotype: poor metabolizers (PMs), intermediate metabolizers (IMs), and extensive metabolizers (EMs). Patients with two null alleles are classified as PMs of drugs that are metabolized by CYP2D6 [6–8]. The major null alleles that cause the PMs in Caucasians are *CYP2D6**3, *CYP2D6**4, *CYP2D6**5, and *CYP2D6**6, which account for nearly 95% of the PMs [9]. Patients classified into PMs in Caucasian or patients with homozygous alleles for low or null enzyme activity in Asian have been reported to have lower plasma endoxifen levels and poorer clinical outcomes when treated with tamoxifen [10–13]. Yet, the mechanism that causes the interindividual differences in responsiveness to tamoxifen is not fully clarified even considering that the effects from genetic variation of *CYP2D6* was taken into account. Although there are numbers of genetic or environmental factors that may affect the response level to tamoxifen, *CYP2D6* genotyping could be the most promising clinical genetic test for predicting the tamoxifen efficacy, and will potentially provide better options to physicians for more optimal hormonal therapy to treat the hormone receptor–positive breast cancer.

13.2.1 Metabolic Pathway of Tamoxifen

Tamoxifen is metabolized in the liver in human body to several metabolites that exhibit a range of pharmacologic activities [4, 14]. Hence, different systemic exposure of those active metabolites can result in a variety of responses to tamoxifen in patients with breast cancer. Tamoxifen can be considered a prodrug.

The parental drug, tamoxifen, has weak affinity to ER by itself, but it is biotransformed into active and inactive metabolites. CYP450 enzymes including CYP2B6, CYP2C9, CYP2C19, CYP2D6, and CYP3A are involved in the biotransformation process for active metabolites of tamoxifen (Fig. 13.1). *N*-demethyl tamoxifen, the most abundant tamoxifen metabolite, is generated from tamoxifen by N-demethylation with CYP3A. Furthermore, endoxifen, one of the active metabolites of tamoxifen, is formed predominantly by the CYP2D6-mediated oxidation of *N*-demethyl tamoxifen as shown in figure 1 [15]. Endoxifen has high affinity for ER and a greater potency equivalent to the potency of 4-hydroxytamoxifen in vitro. Endoxifen reaches the plasma concentrations sixfold or more than 4-OH tamoxifen in patients having tamoxifen [4]. 4-OH tamoxifen and endoxifen undergo conjugation by sulfotransferases (SULTs) and/or Uridine diphosphate-glucuronosyltransferase (UGTs) (Fig. 13.1).

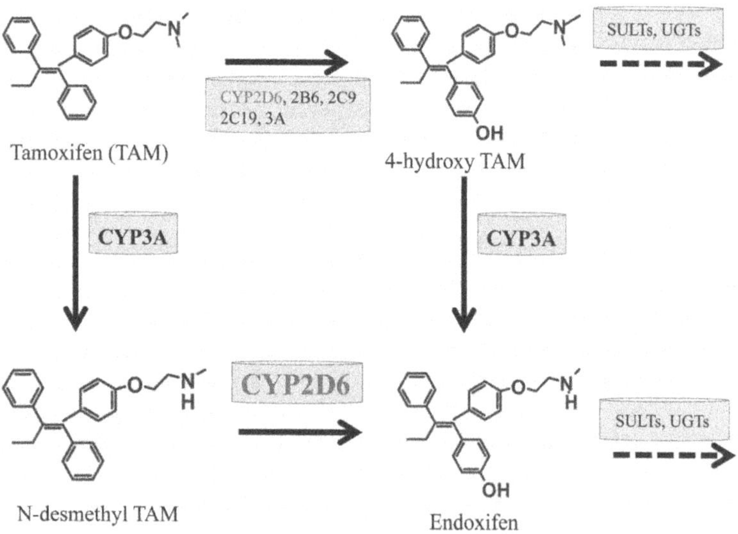

Figure 13.1 Schematic representation of the metabolism of tamoxifen by the cytochrome P450 (CYP) system. The hydroxylated metabolites undergo conjugation by sulfotransferases (SULTs) and uridine diphosphate–glucuronosyltransferases (UGTs).

13.2.2 *CYP2D6* Genotype and Pharmacokinetics of Tamoxifen

Breast cancer patients that are heterozygous or homozygous for decreased- or null-function alleles of *CYP2D6* are reported to show lower plasma concentrations of endoxifen and 4-hydroxytamoxifen compared to the patients with homozygous-wild-type allele [16]. The insufficient plasma concentration of the active metabolite is suggested to be associated with worse clinical outcome in tamoxifen therapy. There is a tamoxifen dose adjustment experiment observing 98 Japanese breast cancer patients that had been taking 20 mg of tamoxifen daily as adjuvant setting [13]. In this study, the tamoxifen dose is increased to 30 and 40 mg per day for the patients who have only one or no normal allele of *CYP2D6*, respectively. The patients with the increased dose show higher plasma endoxifen level compared to those without [13], which is a similar level to the plasma endoxifen concentration from the patients possessing *CYP2D6* wild type receiving 20 mg/day of tamoxifen. This result suggests that dose adjustment can be useful for the patients carrying decreased- or null-function allele of *CYP2D6* to achieve the effective plasma endoxifen level. Another similar genotype-guided tamoxifen dosing study is reported observing 190 patients treated with 20 mg/day of tamoxifen for 4 months [8, 17]. Median concentration of the basal endoxifen is higher in EM compared with either IM or PM patients in the study. The endoxifen concentration rises significantly when the dose is increased from 20 mg to 40 mg in IM and PM patients. There is no longer a significant difference in endoxifen concentrations between EM and IM patients after the dose increase. This study well demonstrates the feasibility of genotype-driven tamoxifen dosing demonstrating that doubling the tamoxifen dose can increase endoxifen concentrations in IM and PM patients [8].

13.2.3 *CYP2D6* Genotype and Tamoxifen Efficacy

There is a major interindividual and ethnic variability in the drug metabolisms by CYP2D6 that can be explained largely by genetic polymorphisms affecting the enzyme activity and its expression level [18]. Some of these alleles are associated with decreased enzymatic activity (e.g., *CYP2D6**10, *17, *41) or with no enzymatic activity (e.g., *3, *4, *5, *6). It has been hyposthesized that women with reduced CYP2D6 enzyme activity, thus presumably low plasma endoxifen concentrations, may have bad prognosis after tamoxifen therapy; however, this hypothesis has not been definitively proven [19].

One of the first studies reported in 2007 by researchers in Mayo Clinic determined *CYP2D6* genotype by extraction of DNA from postmenopausal women treated with 5-year tamoxifen [10]. They report that carriers of a *CYP2D6**4 (null-function allele) has significantly shorter relapse-free survival than those in EMs [10]. In 2010, another *CYP2D6*-tamoxifen study also reported that *CYP2D6* variants causing reduced or null enzymatic activity are significantly associated with shorter recurrence-free survival observed from 282 Japanese patients that received adjuvant tamoxifen monotherapy [12]. In addition to these trials, many clinical studies report the significant association of *CYP2D6* genotype with clinical outcome of patients treated with tamoxifen in adjuvant or metastatic setting [20–26].

13.3 The Controversy in Tamoxifen-*CYP2D6* Study

Several studies report discordant results. In 2012, two large studies reported that the relationship has not been confirmed [27, 28] and that some authors recommend not to use *CYP2D6* test routinely until there is the robust confirmatory data available from adequately powered prospective trials [29, 30]. On the other hand, several reviews point out some common critical issues in *CYP2D6*-tamoxifen pharmacogenomics studies [31–34]. The quality of genotype data, which could be influenced by the accuracy of genotyping methods, coverage of genotyped alleles, and DNA source, is one of the most important issues in the pharmacogenomics study including *CYP2D6*-tamoxifen study [33]. In DNA source, potential misclassification of germline *CYP2D6* genotypes exist in the studies using tumor DNA because the loss of heterozygosity at the *CYP2D6* locus is common in breast cancer [31]. Furthermore, selecting the right participants for the study is another critical issue. It is clear that the breast cancer patients treated with tamoxifen monotherapy are the appropriate participants for *CYP2D6*-tamoxifen study.

13.4 Future Direction of Tamoxifen Pharmacogenomics

In the recent meta-analysis result of data from 4973 tamoxifen-treated patients, the International Tamoxifen Pharmacogenomics Consortium (12 globally distributed sites) reports that CYP2D6 poor metabolizer status is associated with poorer invasive disease-free survival (IDFS) using strict inclusion criteria [37]. The potential role of *CYP2D6* genotype assessment in determining if the patients with ER-positive breast cancer should receive tamoxifen is still controversial. However, some experts agree that *CYP2D6* test may be useful for personalized tamoxifen therapy after robust confirmatory data are available from adequately powered prospective trials [29, 30]. Personalized tamoxifen therapy can improve the treatment efficacy in patients with breast cancer. *CYP2D6* genotype test may be more practical than plasma tamoxifen/endoxifen monitoring because CYP2D6 metabolizer status of a patient can be identified easily and rapidly even without starting the tamoxifen therapy. Recent *CYP2D6*-tamoxifen studies suggest that genotype-guided tamoxifen administration may be beneficial for breast cancer patients in terms of optimizing the treatment of breast cancer.

13.4.1 Pharmacogenomics for Irinotecan: *UGT1A1*

Irinotecan, also known as CPT-11, is a semisynthetic analog of camptothecin that is a natural alkaloid extract of plants [38]. As a topoisomerase 1 inhibitor, irinotecan has been widely used for the treatment of colorectal, non–small-cell lung, and some other cancers [39, 40]. However, it often causes toxicities such as severe myelosuppression and diarrhea. Irinotecan is activated by hydrolysis to SN-38, which is a strong topoisomerase I inhibitor

[41, 42]. UDP-glucuronosyltransferase 1A1 (UGT1A1) catalyzes the glucuronidation of SN-38 (SN-38G; inactive form of SN-38) [43]. The transcriptional efficiency of *UGT1A1* has been inversely correlated to the number of TA repeats in promoter region of this gene, and TA7/TA7 genotype has been known as homozygosity for the *UGT1A1**28. Many reports provide definitive evidence that homozygous for the *UGT1A1**28 allele may experience significant toxicities of irinotecan-containing regimens [44]. On the basis of these evidences, it has been suggested that initial dose of irinotecan must be reduced for patients with homozygous for the *UGT1A1**28 allele [45]. *UGT1A1* pharmacogenetic test has been covered by health insurance in many countries.

13.4.2 Pharmacogenetic Test of *CYP2C9* and *VKORC1* Genotypes for Warfarin Treatment

Warfarin is the most commonly used oral antithrombic in the world. However, adjustment of appropriate dose for each patient is sometimes difficult because of the large interindividual variation in the requirement of this drug. Insufficient dose of warfarin will fail to prevent thrombosis. On the other hand, overdosing will increase the risk of unexpected bleeding. Warfarin is metabolized to the 7- hydroxylated form in humans mainly by cytochrome P450, subfamily 2C, polypeptide 9 (CYP2C9) [46]. Moreover, vitamin K plays an important role in making blood-clotting proteins, and warfarin plays its anticoagulant effect by suppressing the regeneration of vitamin K. The vitamin K epoxide reductase is encoded by vitamin K epoxide reductase complex subunit 1 (*VKORC1*) [47–49]. *CYP2C9* and *VKORC1* have been associated with this interindividual variation of the appropriate dose of warfarin leading to genotype-guided dosing tables in warfarin labeling. To improve the dose adjustment of the drug, FDA declared changes in warfarin labeling in 2007 and publicly issued the dosing tables in 2010 [50]. However, some physicians claim that there must be more elaboration on how to use the two-genotype information in prescribing warfarin. Hence, the guidelines for using the pharmacogenetic information are now expected to improve to maturely establish personalized antithrombic therapy [51].

References

1. Early Breast Cancer Trialists' Collaborative Group, Davies C, Godwin J, et al. Relevance of breast cancer hormone receptors and other factors to the efficacy of adjuvant tamoxifen: patient-level meta-analysis of randomised trials. *Lancet* **378**(9793), 771–784 (2011).
2. Davies C, Pan H, Godwin J, et al. Long-term effects of continuing adjuvant tamoxifen to 10 years versus stopping at 5 years after diagnosis of oestrogen receptor-positive breast cancer: ATLAS, a randomised trial. *Lancet* **381**(9869), 805–816 (2013).
3. Borgna JL, and Rochefort H. Hydroxylated metabolites of tamoxifen are formed in vivo and bound to estrogen receptor in target tissues. *J Biol Chem* **256**(2), 859–868 (1981).
4. Stearns V, Johnson MD, Rae JM, et al. Active tamoxifen metabolite plasma concentrations after coadministration of tamoxifen and the selective serotonin reuptake inhibitor paroxetine. *J Natl Cancer Inst* **95**(23), 1758–1764 (2003).
5. Clarke R, Liu MC, Bouker KB, et al. Antiestrogen resistance in breast cancer and the role of estrogen receptor signaling. *Oncogene* **22**(47), 7316–7339 (2003).
6. Sachse C, Brockmoller J, Bauer S, and Roots I. Cytochrome P450 2D6 variants in a Caucasian population: allele frequencies and phenotypic consequences. *Am J Hum Genet* **60**(2), 284–295 (1997).
7. Griese EU, Zanger UM, Brudermanns U, et al. Assessment of the predictive power of genotypes for the in-vivo catalytic function of CYP2D6 in a German population. *Pharmacogenetics* **8**(1), 15–26 (1998).
8. Irvin WJ, Jr., Walko CM, Weck KE, et al. Genotype-guided tamoxifen dosing increases active metabolite exposure in women with reduced CYP2D6 metabolism: a multicenter study. *J Clin Oncol* **29**(24), 3232–3239 (2011).
9. Broly F, Gaedigk A, Heim M, Eichelbaum M, Morike K, and Meyer UA. Debrisoquine/sparteine hydroxylation genotype and phenotype: analysis of common mutations and alleles of CYP2D6 in a European population. *DNA Cell Biol* **10**(8), 545–558 (1991).
10. Goetz MP, Knox SK, Suman VJ, et al. The impact of cytochrome P450 2D6 metabolism in women receiving adjuvant tamoxifen. *Breast Cancer Res Treat* **101**(1), 113–121 (2007).
11. Lim HS, Ju Lee H, Seok Lee K, Sook Lee E, Jang IJ, and Ro J. Clinical implications of CYP2D6 genotypes predictive of tamoxifen

pharmacokinetics in metastatic breast cancer. *J Clin Oncol* **25**(25), 3837–3845 (2007).

12. Kiyotani K, Mushiroda T, Imamura CK, et al. Significant effect of polymorphisms in CYP2D6 and ABCC2 on clinical outcomes of adjuvant tamoxifen therapy for breast cancer patients. *J Clin Oncol* **28**(8), 1287–1293 (1287).
13. Kiyotani K, Mushiroda T, Imamura CK, et al. Dose-adjustment study of tamoxifen based on CYP2D6 genotypes in Japanese breast cancer patients. *Breast Cancer Res Treat* **131**(1), 137–145 (2011).
14. Ingle JN, Suman VJ, Johnson PA, et al. Evaluation of tamoxifen plus letrozole with assessment of pharmacokinetic interaction in postmenopausal women with metastatic breast cancer. *Clin Cancer Res* **5**(7), 1642–1649 (1999).
15. Desta Z, Ward BA, Soukhova NV, and Flockhart DA. Comprehensive evaluation of tamoxifen sequential biotransformation by the human cytochrome P450 system in vitro: prominent roles for CYP3A and CYP2D6. *J Pharmacol Exp Ther* **310**(3), 1062–1075 (2004).
16. Jin Y, Desta Z, Stearns V, et al. CYP2D6 genotype, antidepressant use, and tamoxifen metabolism during adjuvant breast cancer treatment. *J Natl Cancer Inst* **97**(1), 30–39 (2005).
17. Martinez De Duenas E, Ochoa Aranda E, Blancas Lopez-Barajas I, et al. Adjusting the dose of tamoxifen in patients with early breast cancer and CYP2D6 poor metabolizer phenotype. *Breast* **23**(4), 400–406 (2014).
18. Sistonen J, Sajantila A, Lao O, Corander J, Barbujani G, and Fuselli S. CYP2D6 worldwide genetic variation shows high frequency of altered activity variants and no continental structure. *Pharmacogenet Genomics* **17**(2), 93–101 (2007).
19. Higgins MJ, and Stearns V. CYP2D6 polymorphisms and tamoxifen metabolism: clinical relevance. *Curr Oncol Rep* **12**(1), 7–15 (2010).
20. Schroth W, Antoniadou L, Fritz P, et al. Breast cancer treatment outcome with adjuvant tamoxifen relative to patient CYP2D6 and CYP2C19 genotypes. *J Clin Oncol* **25**(33), 5187–5193 (2007).
21. Schroth W, Goetz MP, Hamann U, et al. Association between CYP2D6 polymorphisms and outcomes among women with early stage breast cancer treated with tamoxifen. *Jama* **302**(13), 1429–1436 (2009).
22. Schroth W, Hamann U, Fasching PA, et al. CYP2D6 polymorphisms as predictors of outcome in breast cancer patients treated with tamoxifen: expanded polymorphism coverage improves risk stratification. *lin Cancer Res* **16**(17), 4468–4477 (2010).

23. Xu Y, Sun Y, Yao L, et al. Association between CYP2D6 *10 genotype and survival of breast cancer patients receiving tamoxifen treatment. *Ann Oncol* **19**(8), 1423–1429 (2008).

24. Goetz MP, Suman VJ, Hoskin TL, et al. CYP2D6 metabolism and patient outcome in the Austrian Breast and Colorectal Cancer Study Group trial (ABCSG) 8. *Clin Cancer Res* **19**(2), 500–507 (2013).

25. Teh LK, Mohamed NI, Salleh MZ, et al. The risk of recurrence in breast cancer patients treated with tamoxifen: polymorphisms of CYP2D6 and ABCB1. *AAPS J* **14**(1), 52–59 (2012).

26. Sirachainan E, Jaruhathai S, Trachu N, et al. CYP2D6 polymorphisms influence the efficacy of adjuvant tamoxifen in Thai breast cancer patients. *Pharmgenomics Pers Med* **5**, 149–153 (2012).

27. Regan MM, Leyland-Jones B, Bouzyk M, et al. CYP2D6 genotype and tamoxifen response in postmenopausal women with endocrine-responsive breast cancer: the breast international group 1-98 trial. *J Natl Cancer Inst* **104**(6), 441–451 (2012).

28. Rae JM, Drury S, Hayes DF, et al. CYP2D6 and UGT2B7 genotype and risk of recurrence in tamoxifen-treated breast cancer patients. *J Natl Cancer Inst* **104**(6), 452–460 (2012).

29. Hoskins JM, Carey LA, and McLeod HL. CYP2D6 and tamoxifen: DNA matters in breast cancer. *Nat Rev Cancer* **9**(8), 576–586 (2009).

30. Lash TL, Cronin-Fenton D, Ahern TP, et al. CYP2D6 inhibition and breast cancer recurrence in a population-based study in Denmark. *J Natl Cancer Inst* **103**(6), 489–500 (2011).

31. Goetz MP, Sun JX, Suman VJ, et al. Loss of heterozygosity at the CYP2D6 locus in breast cancer: implications for germline pharmacogenetic studies. *J Natl Cancer Inst* **107**(2), dju401 (2015).

32. Nakamura Y, Ratain MJ, Cox NJ, Mcleod HL, Kroetz DL, and Flockhart DA. Re: CYP2D6 genotype and tamoxifen response in postmenopausal women with endocrine-responsive breast cancer: the Breast International Group 1-98 trial. *J Natl Cancer Inst* **104**(16), 1264; author reply 1266–1268 (2012).

33. Kiyotani K, Mushiroda T, Zembutsu H, and Nakamura Y. Important and critical scientific aspects in pharmacogenomics analysis: lessons from controversial results of tamoxifen and CYP2D6 studies. *J Hum Genet* **58**(6), 327–333 (2013).

34. Ratain MJ, Nakamura Y, and Cox NJ. CYP2D6 genotype and tamoxifen activity: understanding interstudy variability in methodological quality. *Clin Pharmacol Ther* **94**(2), 185–187 (2013).

35. Nowell SA, Ahn J, Rae JM, et al. Association of genetic variation in tamoxifen-metabolizing enzymes with overall survival and recurrence of disease in breast cancer patients. *Breast Cancer Res Treat* **91**(3), 249–258 (2005).
36. Abraham JE, Maranian MJ, Driver KE, et al. CYP2D6 gene variants: association with breast cancer specific survival in a cohort of breast cancer patients from the United Kingdom treated with adjuvant tamoxifen. *Breast Cancer Res* **12**(4), R64 (2010).
37. Province MA, Goetz MP, Brauch H, et al. CYP2D6 genotype and adjuvant tamoxifen: meta-analysis of heterogeneous study populations. *Clin Pharmacol Ther* **95**(2), 216–227 (2013).
38. Rothenberg ML. Irinotecan (CPT-11): recent developments and future directions–colorectal cancerand beyond. *Oncologist* **6**(1), 66–80 (2001).
39. Vanhoefer U, Harstrick A, Achterrath W, et al. Irinotecan in the treatment of colorectal cancer: clinical overview. *J Clin Oncol* **19**(5), 1501–1518 (2001).
40. Cha PC, Mushiroda T, Zembutsu H, et al. Single nucleotide polymorphism in ABCG2 is associated with irinotecan-induced severe myelosuppression. *J Hum Genet* **54**(10), 572–580 (2009).
41. Innocenti F, Undevia SD, Iyer L, et al. Genetic variants in the UDP-glucuronosyltransferase 1A1 gene predict the risk of severe neutropenia of irinotecan. *J Clin Oncol* **22**(8), 1382–1388 (2004).
42. Kawato Y, Aonuma M, Hirota Y, et al. Intracellular roles of SN-38, a metabolite of the camptothecin derivative CPT-11, in the antitumor effect of CPT-11. *Cancer Res* **51**(16), 4187–4191 (1991).
43. Iyer L, King CD, Whitington PF, et al. Genetic predisposition to the metabolism of irinotecan (CPT-11). Role of uridine diphosphate glucuronosyltransferase isoform 1A1 in the glucuronidation of its active metabolite (SN-38) in human liver microsomes. *Clin Invest* **101**(4), 847–854 (1998).
44. Crona DJ, Ramirez J, Qiao W, et al. Clinical validity of new genetic biomarkers of irinotecan neutropenia: an independent replication study. *Pharmacogenomics J* 16(1), 54–59 (2016).
45. United States Food and Drug Administration: Camptosar label. http://www.fda.gov/cder/foi/label/2005/020571s024,027,028lbl.pdf
46. Rettie AE, Korzekwa KR, Kunze KL, et al. Hydroxylation of warfarin by human cDNA-expressed cytochrome P-450: a role for P-4502C9 in the etiology of (S)-warfarin-drug interactions. *Chem Res Toxicol* **5**(1), 54–59 (1992).

47. Rost S, Fregin A, Ivaskevicius V, et al. Mutations in VKORC1 cause warfarin resistance and multiple coagulation factor deficiency type 2. *Nature* **427**(6974), 537–541 (2004).

48. Li T, Chang CY, Jin DY, et al. Identification of the gene for vitamin K epoxide reductase. *Nature* **427**(6974), 541–544 (2004).

49. Mushiroda T, Ohnishi Y, Saito S, et al. Association of VKORC1 and CYP2C9 polymorphisms with warfarin dose requirements in Japanese patients. *J Hum Genet* **51**(3), 249–253 (2006).

50. FDA Approves Updated Warfarin (Coumadin) Prescribing Information (2007).

51. Jorgensen AL, FitzGerald RJ, Oyee J, et al. Influence of CYP2C9 and VKORC1 on patient response to warfarin: a systematic review and meta-analysis. *PLoS One* **7**(8), e44064 (2012).

Index